Torsten Hein

Regularization in Banach spaces – convergence rates theory

HABILITATIONSSCHRIFT

zur Erlangung des akademischen Grades
Doctor rerum naturalium habilitatus
(Dr. rer. nat. habil.)

TECHNISCHE UNIVERSITÄT CHEMNITZ
Fakultät für Mathematik

CHEMNITZ UNIVERSITY
OF TECHNOLOGY

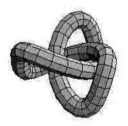

Bibliografische Information der Deutschen Nationalbibliothek

Die Deutsche Nationalbibliothek verzeichnet diese Publikation in der Deutschen Nationalbibliografie; detaillierte bibliografische Daten sind im Internet über http://dnb.d-nb.de abrufbar.

ISBN 978-3-8325-2745-7

Logos Verlag Berlin GmbH
Comeniushof, Gubener Str. 47,
10243 Berlin
Tel.: +49 (0)30 42 85 10 90
Fax: +49 (0)30 42 85 10 92
INTERNET: http://www.logos-verlag.de

Contents

List of Tables

List of Figures

Symbols

Sets and spaces

$\mathbb{N}$, $\mathbb{R}$	natural/real numbers
$\mathcal{X}$, $\mathcal{Y}$	real Banach spaces
$\mathcal{X}^*$, $\mathcal{Y}^*$	dual spaces of $\mathcal{X}$ and $\mathcal{Y}$
$\mathcal{N}(A), \mathcal{R}(A)$	kernel/range of the linear operator A
$\mathcal{D}(F)$	domain of the forward operator F
$\mathcal{D}(P)$	domain of the penalty functional P
$\mathcal{D}$	domain of the Tikhonov functional T_{α,y^δ}
$\mathcal{D}_B$	Bregman domain of P
$\partial P(x)$	subdifferential of P at x
$\mathcal{M}_L(\cdot,\cdot)$	sets of elements satisfying the low order approximate source condition
$d_L(\cdot,\xi)$	distance function of ξ with respect to the low order reference source condition
$\mathcal{M}_H(\cdot,\cdot)$	sets of elements satisfying the high order approximate source condition
$d_p(\cdot,\xi)$	distance function of ξ with respect to the high order reference source condition and parameter p
$\mathcal{M}$	range of the operator $x \mapsto A^\star J_p(A\,x)$

Operators and functionals

$\|\cdot\|$, $\langle\cdot,\cdot\rangle$, $\langle\cdot,\cdot\rangle_{\mathcal{X}}$	norm, duality product. scalar product of the Hilbert space $\mathcal{X}$
$F(\cdot)$	nonlinear operator of the forward problem
A	linear operator of the forward problem
$P(\cdot)$	penalty functional
A^*, $A^\star$	Hilbert space adjoint/dual operator of A
P_h, Q_h	projection operators onto finite-dimensional subspaces of $\mathcal{X}/\mathcal{Y}$
A_h	discretization of the operator A
$F'(x)$	derivative of F at x
$T_{\alpha,y^\delta}(\cdot)$	Tikhonov functional with data y^δ and regularization parameter α
$\mathcal{D}_\xi(\cdot,x)$	Bregman distance of P at x and $\xi \in \partial P(x)$
$D_{Y,p}(\cdot,y)$	Bregman distance of $y \mapsto p^{-1}\|y\|^p$ in $\mathcal{Y}$
$\Delta_p(\cdot,x)$,$\Delta^*_{p^*}(\cdot,x^*)$	Bregman distances of $x \mapsto p^{-1}\|x\|^p$ and $x^* \mapsto (p^*)^{-1}\|x^*\|^{p^*}$ in $\mathcal{X}$ respectively $\mathcal{X}^*$
$J_p(\cdot)$, $J^*_{p^*}(\cdot)$	duality mapping with gauge function $t \mapsto t^{p-1}$ in $\mathcal{X}$ or $\mathcal{Y}$ and $t \mapsto t^{p^*-1}$ in $\mathcal{X}^*$ or $\mathcal{Y}^*$

Elements

$x^\dagger$, $\xi^\dagger$	exact solution, element of the subdifferential $\partial P(x^\dagger)$

x_α^δ	Tikhonov regularized approximate solution
$x_{\alpha,h}^\delta$	Tikhonov regularized approximate discretized solution
$x_{N(\delta,y^\delta)}$	iteratively regularized approximate solution
y, y^δ	exact/noisy data
ψ_n^*	gradient of $x \mapsto p^{-1}\|F(x) - y^\delta\|^p$ at the n-th iterate x_n^δ
Ψ_n^*	gradient of the Tikhonov functional at the n-th iterate x_n

Chapter 1

Introduction

Mathematical models in natural, medical and economic sciences try to describe relations between *causes* x and corresponding *effects* y. Many of such models can be formulated as operator equations

$$F(x) = y, \qquad x \in \mathcal{D}(F) \subseteq \mathcal{X},\ y \in \mathcal{Y}, \tag{1.1}$$

where $F : \mathcal{D}(F) \subseteq \mathcal{X} \longrightarrow \mathcal{Y}$ denotes a mapping between the *pre-image* or *solution space* $\mathcal{X}$ and the *image space* $\mathcal{Y}$. The domain $\mathcal{D}(F)$ describes the set of all admissible causes x. If F linear and bounded and $\mathcal{D}(F) = \mathcal{X}$ then we write $F = A$ and the nonlinear equation (1.1) is relaced by the linear equation

$$Ax = y, \qquad x \in \mathcal{X},\ y \in \mathcal{Y}. \tag{1.2}$$

Calculating y from given cause x is called the *forward problem*. However, many problems in practical applications are of inverse nature: one is interested in the quantity x which cannot be observed or measured directly. On the other hand it is possible to obtain data in form of the corresponding variable y. In addition to the higher numerical effort of solving such *inverse problems* we have to take into account the effect of *ill-posedness*, i.e for given $y \in \mathcal{Y}$

- a solution x of equation (1.1) might not exist,
- such solution might not be unique or
- the solution might not depend stable on the data y.

The last point is the most crucial one in the numerical treatment of inverse problems. Based on measurement and/or discretization errors the exact value y is usually unknown. Only *noisy* data y^δ is given where $\delta > 0$ describes an estimate of the quantity of the noise. We introduce the notion of convergence in $\mathcal{X}$ and $\mathcal{Y}$ and assume now $y^\delta \to y$ as $\delta \to 0$. Even if unique elements x^δ and x exist satisfying $F(x^\delta) = y^\delta$ and $F(x) = y$ one cannot ensure $x^\delta \to x$ as $\delta \to 0$ when the problem is unstable.

Therefore, *regularization methods* were developed which should overcome the problem of ill-posedness. Introduced in [79] and [92], *Tikhonov regularization* is certainly the most prominent and deepest studied example of such methods: for given parameter $1 \leq p < \infty$, regularization parameter $\alpha > 0$ and convex stabilizing functional $P : \mathcal{D}(P) \subseteq \mathcal{X} \longrightarrow \mathbb{R}$ with domain $\mathcal{D}(P)$ we calculate the solution x_α^δ of the minimizing problem

$$T_{\alpha,y^\delta}(x) := \frac{1}{p}\|F(x) - y^\delta\|^p + \alpha\, P(x) \to \min \quad \text{subject to} \quad x \in \mathcal{D}(F) \cap \mathcal{D}(P). \tag{1.3}$$

The functional $T_{\alpha,y^\delta} : \mathcal{D}(F) \cap \mathcal{D}(P) \subseteq \mathcal{X} \longrightarrow \mathbb{R}$ is referred to as the underlying *Tikhonov functional* and x_α^δ is called the *(Tikhonov-)regularized approximate solution* of equation (1.1). Then in the context of regularization theory we have to deal with the following points:

(i) **Well-posedness.** For each $\alpha > 0$ and each data $y^\delta \in \mathcal{Y}$ there exists a (unique) minimizer x_α^δ of the problem (1.3) which depends stable on the given data.

(ii) **Convergence.** For $\delta \to 0$ and proper choice $\alpha = \alpha(\delta) \to 0$ we have convergence $x_\alpha^\delta \to x$ with $F(x) = y$.

(iii) **Convergence rates.** Which speed or rate of convergence $x_\alpha^\delta \to x$ can be (maximally) achieved and how the parameter α must be chosen to obtain this rate?

There exists a vast number of literature about regularization theory when $\mathcal{X}$ and $\mathcal{Y}$ are supposed to be Hilbert spaces, see e.g. [93], [7], [72], [21] and [84]. However, practical applications necessitate also the regard of Banach spaces in regularization theory. In particular, *bounded* or *total variation penalization* and *maximum entropy regularization* as special variants of (1.3) were successfully applied in practice, see [1] and [19]. For their analytical treatment the choice of $\mathcal{X}$ as a convenient Banach space is essential. A full Banach space setting was considered in [89] where well-posedness and convergence of Tikhonov regularization was studied for a wide class of penalty functionals. The results concerning convergence rates in Banach spaces are much more limited. The probably first result can be found in [19] for maximum entropy regularization whereas in the paper [12] an idea for formulating convergence rates for general penalty functionals was presented. In both papers the space $\mathcal{Y}$ was still assumed to be a Hilbert space. The first convergence rates in a full Banach space setting can be found in [81] for linear and in [45] for nonlinear operator equations. All four results base on a rather strong smoothness assumption on the exact solution x of (1.1). A general approach for presenting a convergence rates theory under weaker smoothness of the exact solution was introduced in [33] for $p = 1$ and [36] for $1 < p < \infty$ for linear problems as well as nonlinear problems satisfying a certain non-linearity restriction. Later on, in [40] these results were extended to more general classes of nonlinear equations. One central goal of this thesis is the presentation of this theory in a unified framework and show their relation to known convergence results in Hilbert spaces.

Consequently, the thesis is structured in the following way:

- The rest of this chapter deals with two motivating examples – one arises of practical applications, the other one is of theoretical nature – for deriving a generalized convergence rates theory for Tikhonov regularization in Banach spaces.

- In Chapter 2 we present the mathematical basis of our investigations. In doing so we first recall geometric properties of Banach spaces such as *smoothness* and *convexity* and their correlations. Furthermore, *duality mappings* as well as *Bregman distances* of convex functionals were introduced and important properties specified which will be used later in our analysis. The chapter closes with a comparison of the role of *(general) source conditions* and *approximate source conditions* in classical regularization theory for linear operator equations in Hilbert spaces.

- Chapter 3 deals with the theory of Tikhonov regularization in Banach spaces. Introducing basic assumptions we shortly recall (known) existence, stability and convergence results in the first section. The next two sections deeply study the convergence rates analysis of Tikhonov regularization. Based on two different type of approximate source conditions we present a various number of convergence rates results taking into account different convexity assumptions for the penalty term as well as smoothness properties of the underlying Tikhonov functional. On the one hand we present a *low order* convergence rates analysis which does not require any smoothness of the Tikhonov functional. However, under certain smoothness information about the (exact) solution of (1.1) respectively (1.2) higher convergence rates might be proved. This is done in the *high order* convergence rates analysis needing additional smoothness properties of the Tikhonov functional. Furthermore we shortly discuss methods for finding minimizers of Tikhonov functionals in a simplified situation. Afterwards we present discretization strategies for the numerical solution which try to avoid too fine discretization on the one and do not spoil the convergence rates result of the previous section on the other side. Finally, numerical case studies were presented which should compare theoretical with numerical results.

- The last chapter is devoted to iterative regularization methods. In particular, accelerated Landweber iteration and an accelerated modified Landweber iteration were taken into account. For both cases convergence and stability results were derived whereas in the latter case we also prove a convergence rates result. Again, numerical examples should illustrate these theoretical considerations.

In particular, it should be demonstrated that by the approach of approximate source conditions the convergence rates theory of regularization methods in Hilbert spaces can be transferred in wide parts in a very natural way into a Banach space theory.

1.1 Motivation I – Exact penalization

Inspired by the fundamental paper [12] we analyze the following situation:

- We deal with the linear ill-posed operator equation (1.2) and $\mathcal{X}$ and $\mathcal{Y}$ denote real Hilbert spaces.
- The element $x^\dagger \in \mathcal{X}$ denotes the unique minimum-norm-solution of equation (1.2), i.e. $x^\dagger = \operatorname{argmin}\{\|x\| \,:\, A\,x = y\}$. In particular, we assume that equation (1.2) has a solution.
- We suppose given noisy data $y^\delta \in \mathcal{Y}$ with noise $\|y^\delta - y\| \le \delta$ for some $\delta > 0$. Instead of classical Tikhonov regularization we deal with a so-called exact-penalization-

model with non-squared residual term, i.e. we solve the non-quadratic minimizing problem

$$\|Ax - y^\delta\| + \frac{\alpha}{2}\|x\|^2 \to \min \quad \text{subject to} \quad x \in \mathcal{X}. \tag{1.4}$$

Then the following statements holds:

(i) For each $\alpha > 0$ there exists a unique solution $x_\alpha^\delta \in \mathcal{X}$ of (1.4) depending stable on the given date y^δ. The uniqueness is clear because of the strict convexity of the functional in (1.4). The existence and stability proof can be carried out by standard techniques, see e.g. [1], [22] and [89] in more general situations.

(ii) If the element $x^\dagger$ additionally satisfies the source condition

$$x^\dagger = A^*\omega, \qquad \omega \in \mathcal{Y} \tag{1.5}$$

and the regularization parameter $\alpha > 0$ is chosen such that $\alpha \le \|\omega\|$, then

$$\|x_\alpha^\delta - x^\dagger\| = \mathcal{O}\left(\sqrt{\delta}\right) \quad \text{as} \quad \delta \to 0. \tag{1.6}$$

This is an immediate consequence of [12, Theorem 5]. Here, $A^* : \mathcal{Y} \longrightarrow \mathcal{X}$ denotes the corresponding Hilbert space adjoint operator of A.

The convergence rate result (1.6) seems to be quite surprising at the first moment. On the one hand, the suggested parameter choice stays in opposite to classical parameter choices where we have $\alpha = \alpha(\delta) \to 0$ as $\delta \to 0$. However, the error estimate in [12] indicates that any choice $\alpha \to 0$ as $\delta \to 0$ leads to a lower convergence rate than (1.6). Moreover, the optimal choice of the regularization parameter $\alpha > 0$ does not depend on the noise level δ. One might think that this observation contradicts the well-known Bakushinsky-veto [4] which says that no linear regularization method can be found with a choice of the parameter α which does not depend on the noise-level δ. However, even we deal with a linear operator equation, the exact penalization model (1.4) defines a nonlinear regularization method. So the Bakushinsky-veto cannot be applied to the exact-penalization approach (1.4). Furthermore, the convergence rate result (1.6) was established only under the additional source condition (1.5). Presented in a more general situation it was left as open problem in [12] what happens when (1.5) is violated. However, in this specific Hilbert space situation we can easily prove the following convergence rate result by assuming a (lower) power-type source condition. Main tool in the proof is the so-called interpolation inequality, see [21, Section 2.3], which is only applicable for operators mapping between Hilbert spaces. Later on, in Section 2.4 we will present an alternative proof based on distance functions and their correlations to power-type source conditions, see also [33]. As we will see the second variant can transferred without further ado to mappings between Banach spaces which is important for our further analsyis. We present the underlying result.

Lemma 1.1.1 *Let $\mathcal{X}$ and $\mathcal{Y}$ be Hilbert spaces and $x^\dagger = (A^*A)^\mu\omega$ for some $\omega \in \mathcal{X}$ and some $0 < \mu < \frac{1}{2}$. Then a parameter choice $\alpha \sim \delta^{\frac{1+2\mu}{1-2\mu}}$ leads to a convergence rate*

$$\|x_\alpha^\delta - x^\dagger\| = \mathcal{O}\left(\delta^{\frac{2\mu}{2\mu+1}}\right) \quad \textit{as} \quad \delta \to 0. \tag{1.7}$$

PROOF. By the minimizing property of x_α^δ we conclude

$$\|A x_\alpha^\delta - y^\delta\| + \frac{\alpha}{2}\left(\|x_\alpha^\delta\|^2 - \|x^\dagger\|^2\right) \le \|A x^\dagger - y^\delta\| \le \delta.$$

Let $\langle\cdot,\cdot\rangle_{\mathcal{X}}$ and $\langle\cdot,\cdot\rangle_{\mathcal{Y}}$ denote the underlying scalar products in $\mathcal{X}$ and $\mathcal{Y}$ respectively. Then we derive with the source condition that

$$\|x_\alpha^\delta\|^2 - \|x^\dagger\|^2 = \|x_\alpha^\delta - x^\dagger\|^2 + 2\langle x^\dagger, x_\alpha^\delta - x^\dagger\rangle_{\mathcal{X}} = \|x_\alpha^\delta - x^\dagger\|^2 + 2\left\langle \omega, (A^*A)^\mu (x_\alpha^\delta - x^\dagger)\right\rangle_{\mathcal{Y}}.$$

We set $\nu := 2\mu$. Then we conclude from the interpolation inequality, see [21, Section 2.3],

$$\begin{aligned} \|A x_\alpha^\delta - y^\delta\| + \frac{\alpha}{2}\|x_\alpha^\delta - x^\dagger\|^2 &\le \delta + \alpha\|\omega\| \|(A^*A)^{\frac{\nu}{2}}(x_\alpha^\delta - x^\dagger)\| \\ &\le \delta + \alpha\|\omega\| \|A x_\alpha^\delta - y\|^\nu \|x_\alpha^\delta - x^\dagger\|^{1-\nu}. \end{aligned}$$

Applying Young's inequality we obtain

$$\left(\nu^\nu \alpha\|\omega\| \|x_\alpha^\delta - x^\dagger\|^{1-\nu}\right)\left(\nu^{-\nu}\|A x_\alpha^\delta - y\|^\nu\right) \le \|A x_\alpha^\delta - y\| + (1-\nu)\left(\nu^\nu\|\omega\|\right)^{\frac{1}{1-\nu}} \alpha^{\frac{1}{1-\nu}} \|x_\alpha^\delta - x^\dagger\|.$$

With constant $\mathcal{C} := (1-\nu)\left(\nu^\nu\|\omega\|\right)^{\frac{1}{1-\nu}}$ and $\|A x_\alpha^\delta - y\| \le \delta + \|A x_\alpha^\delta - y^\delta\|$ we arrive at

$$\|x_\alpha^\delta - x^\dagger\|^2 \le \frac{4\,\delta}{\alpha} + 2\mathcal{C}\,\alpha^{\frac{1}{1-\nu}-1}\|x_\alpha^\delta - x^\dagger\| = \frac{4\,\delta}{\alpha} + 2\mathcal{C}\,\alpha^{\frac{\nu}{1-\nu}}\|x_\alpha^\delta - x^\dagger\|$$

From the implication

$$a,\, b,\, c > 0,\; a^2 \le b^2 + c\,a \;\Rightarrow\; a \le b + c$$

we get

$$\|x_\alpha^\delta - x^\dagger\| \le 2\frac{\sqrt{\delta}}{\sqrt{\alpha}} + 2\mathcal{C}\,\alpha^{\frac{\nu}{1-\nu}} = 2\frac{\sqrt{\delta}}{\sqrt{\alpha}} + 2\mathcal{C}\,\alpha^{\frac{2\mu}{1-2\mu}}.$$

Balancing both terms in the error estimate we set

$$\delta = \alpha^{\frac{4\mu}{1-2\mu}+1} = \alpha^{\frac{1+2\mu}{1-2\mu}} \;\Leftrightarrow\; \alpha = \delta^{\frac{1-2\mu}{1+2\mu}}$$

and derive

$$\|x_\alpha^\delta - x^\dagger\| \le (2 + 2\mathcal{C})\,\delta^{\frac{2\mu}{2\mu+1}},$$

which proves the accordant convergence rate. ■

We observe two points: first of all, the parameter choice for α now depends on the noise level δ again. In fact, the specific choice of the regularization parameter α in [12] is essentially a consequence of the underlying source condition. Moreover, the convergence rate (1.7) is known to be the optimal one for any regularization method such as Tikhonov regularization under the assumed source condition, see e.g. [21, Setion 5.1]. Hence classical Tikhonov regularization and the exact-penalization approach (1.4) provide the same (optimal) convergence rates. Only the decay rate $\alpha(\delta) \to 0$ as $\delta \to 0$ for the optimal choice of the regularization parameter α differs.

On the other hand, in [12] a more general situation was considered:

- The space $\mathcal{X}$ is assumed to be a real Banach space with dual space $\mathcal{X}^*$.
- The penalty term $\frac{1}{2}\|x\|^2$ is replaced by an arbitrary (proper chosen) convex functional $P : \mathcal{X} \longrightarrow [0,\infty]$ and $x^\dagger \in \mathcal{X}$ denotes a P-minimizing solution of (1.2), i.e. $x^\dagger = \operatorname{argmin}\{P(x) \,:\, A\,x = y\}$. Notice, that if equation (1.2) has a solution, then the same conditions for the penalty functional providing the existence of a minimizer x_α^δ of the Tikhonov functional also ensure the existence of a P-minimizing solution.
- The source condition is formulated for some $\xi^\dagger \in \partial P(x^\dagger) \subset \mathcal{X}^*$ where $\partial P(x^\dagger)$ denotes the subdifferential of P at $x^\dagger$. In particular, the existence of an element $\tilde{\omega} \in \mathcal{Y}$ is assumed such that $\xi^\dagger = A^*\tilde{\omega}$, where $A^* : \mathcal{Y} \longrightarrow \mathcal{X}^*$ is the dual operator in this Banach-/Hilbert space setting. We recall, that the subdifferential $\partial P(x)$ at $x \in \mathcal{X}$ is defined as

$$\partial P(x) := \{\xi \in \mathcal{X}^* \;:\; P(x+h) \geq P(x) + \langle \xi, h\rangle,\; \forall\, h \in \mathcal{X}\}\,.$$

- The authors introduced Bregman distances

$$D_\xi(\tilde{x}, x) := P(\tilde{x}) - P(x) - \langle \xi, \tilde{x} - x\rangle, \quad \tilde{x} \in \mathcal{X},$$

  of P at $x \in \mathcal{X}$ and $\xi \in \partial P(x)$ and the convergence rates result was formulated with respect to the Bregman distance $D_{\xi^\dagger}(x_\alpha^\delta, x^\dagger)$.

The last two points essentially influenced the development of the convergence rates theory for Tikhonov regularization approaches with general (convex) stabilizing functional in Banach spaces. These approaches allow to develop a unified convergence rates theory for a wide class of stabilizing functionals which coincides with the classical results in Hilbert spaces.

However, several questions occur:

1. Does a similar convergence rates result hold when the space $\mathcal{Y}$ is also supposed to be a Banach space? Then the accordant source condition is $\xi^\dagger = A^\star\hat{\omega}$, $\hat{\omega} \in \mathcal{Y}^*$, with the dual operator $A^\star : \mathcal{Y}^* \longrightarrow \mathcal{X}^*$ of A.
2. Can we still prove convergence rates (with respect to Bregman distances or norm) when this source condition is violated? The source condition in Lemma 1.1.1 and the underlying proof of is essentially based on spectral calculus of functions of selfadjoint operators which is a tool in Hilbert spaces. So, if $\mathcal{X}$ and/or $\mathcal{Y}$ are Banach spaces this technique cannot be applied. In particular, a new concept for measuring the violation of the source condition of type (1.5) in Banach spaces is needed.
3. If we have in mind L^p-spaces, $1 \leq p < \infty$, one might consider regularization approaches of the form

$$\frac{1}{p}\|A\,x - y^\delta\|^p + \alpha\, P(x) \to \min \quad \text{subject to} \quad x \in \mathcal{X}.$$

   Can we prove convergence rates for arbitrary exponent p?
4. It is well-known that the convergence rate (1.6) is not a saturation rate for Tikhonov regularization in Hilbert spaces. In particular, we can prove higher convergence rates if the element $x^\dagger$ satisfies a stronger source condition than (1.5). So, can

we also prove higher convergence rates in a Banach space situation? How do the corresponding source conditions look like?

5. Can such convergence rates theory be generalized to nonlinear equations? In [12] one can find a short remark (without proof) that the results of the paper can transferred to nonlinear operator equations if the underlying operator satisfies some nonlinearity restrictions.

These are exactly the questions which should be discussed in this thesis in detail.

1.2 Motivation II – Sparsity reconstruction

The second example arises from observations in practical applications which we will illustrate by a little numerical example. Before we present some numerical results we introduce a linear sample operator which serves as standard example throughout this thesis.

Example 1.2.1 (Standard example) *Let $\mathcal{X}$ and $\mathcal{Y}$ denote Banach spaces of functions on the interval $[0,1]$. We introduce A as the the linear operator of integration, i.e. the sample operator $A:\mathcal{X}\longrightarrow\mathcal{Y}$ given as*

$$[A\,x](t) := \int_0^t x(\tau)\,d\tau, \qquad x\in\mathcal{X}.$$

For the choice $\mathcal{X}=\mathcal{Y}=L^2(0,1)$ this is the standard example for introducing linear ill-posed problems in Hilbert spaces, see e.g. [21, Section 1.1], see also [29]. Because of its simple structure the operator A is well-studied and an ideal candidate for analytical illustrations and numerical experiments, see [25] and [88].

Here we apply this example with $\mathcal{Y}=L^2(0,1)$. The choice of $\mathcal{X}$ we discuss later. We assume that the exact solution $x^\dagger$ has a 'sparse' structure, i.e. $x^\dagger$ is nonzero only on small subintervals of $[0,1]$. Such example is closely related with the problem of *sparsity reconstruction* where $x^\dagger$ is supposed to be an element of a finite-dimensional subspace of $\mathcal{X}$ in some given, countable basis of $\mathcal{X}$, see e.g. [10], [14] and [25].

We apply a Tikhonov regularization approach. Choosing a parameter $q>1$ we here deal with the minimization problem

$$\frac{1}{2}\|A\,x-y^\delta\|^2+\frac{\alpha}{q}\|x\|^q_{L^q}\to\min \quad \text{subject to} \quad x\in\mathcal{X}. \tag{1.8}$$

In particular, for $q\neq 2$ we consider a non-quadratic penalty functional

$$P(x) := \frac{1}{q}\|x\|^q_{L^q} = \frac{1}{q}\int_0^1 |x(t)|^q\,dt, \qquad x\in\mathcal{X}\cap L^q(0,1).$$

We exclude here the choice $q=1$ in order to keep the Tikhonov functional differentiable. The following observation suggests to favor parameters $q>1$ close to one: for given

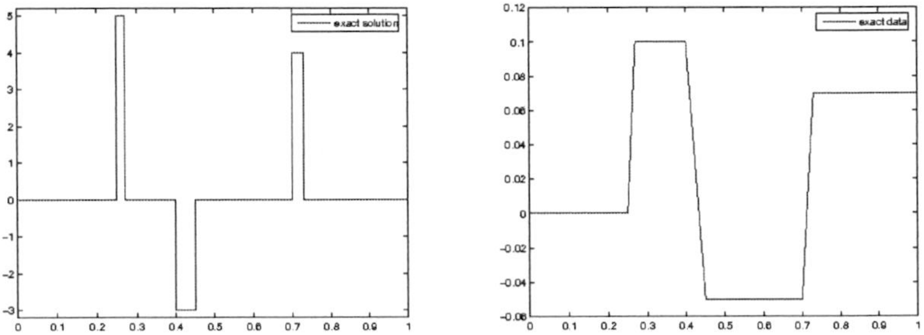

Figure 1.1: Exact solution (left plot) and exact data (right plot)

$x \in \mathcal{X}$ small function values of x are penalized stronger for small q than in the quadratic case $q = 2$. So we can expect that a regularized solution x_α^δ of (1.8) for smaller q has a higher tendency to stay close to zero on the intervals where $x^\dagger$ vanishes than for $q = 2$. Of course this advantageous property of the penalty functional is paid by the price that larger function values are penalized weaker.

Therefore we consider the following piecewise constant sample function

$$x^\dagger(t) := \begin{cases} 5, & t \in [0.25, 0.27], \\ -3, & t \in [0.4, 0.45], \\ 4, & t \in [0.7, 0.73], \\ 0, & \text{else}. \end{cases}$$

In particular, $x^\dagger$ has only three small subintervals where the function does not vanish. The corresponding data $y := Ax^\dagger$ is a piecewise linear and continuous function. Both functions are plotted in Figure 1.1. For discretization we divide the interval $[0,1]$ into $n = 500$ equidistant subintervals. We set $t_j := j/n$, $0 \le j \le n$, and approximate x by n piece-wise constant ansatz functions, i.e.

$$x(t) \approx \sum_{j=1}^{n} x_j \varphi_j(t), \quad t \in [0,1], \quad \text{with } \varphi_j(t) = \chi_{(t_{j-1},t_j]}, \ 1 \le j \le n.$$

If x is supposed to be continuous on the intervals (t_{j-1}, t_j) we can set e.g. $x_j := x((t_{j-1} + t_j)/2)$. Then the specific discretization implies that we have no discretization error for discretizing the exact solution $x^\dagger$. For the discretization of the data $y \in \mathcal{Y}$ we can take the function values of $y \in \mathcal{Y}$ at the end points of the n subintervals, i.e. we approximate

$$y(t) \approx \sum_{j=1}^{n} y_j \varphi_j(t), \quad t \in [0,1], \qquad \text{with } \ y_j := y(t_j), \ 1 \le j \le n.$$

The corresponding discretization of the norms and duality products is induced by the specific choice of the ansatz functions $\varphi_j(t)$, $1 \le j \le n$, $t \in [0,1]$.

	$q = 2$		$q = 1.5$		$q = 1.1$	
δ_{rel}	α_{opt}	$\frac{\|x_\alpha^\delta - x^\dagger\|_q}{\|x^\dagger\|_q}$	α_{opt}	$\frac{\|x_\alpha^\delta - x^\dagger\|_q}{\|x^\dagger\|_q}$	α_{opt}	$\frac{\|x_\alpha^\delta - x^\dagger\|_q}{\|x^\dagger\|_q}$
10^{-1}	$4.1 \cdot 10^{-5}$	0.39398	$6.4 \cdot 10^{-5}$	0.33670	$3.5 \cdot 10^{-4}$	0.34328
10^{-2}	$3.5 \cdot 10^{-6}$	0.19409	$5.5 \cdot 10^{-6}$	0.14203	$7.9 \cdot 10^{-6}$	0.12259
10^{-3}	$2.1 \cdot 10^{-8}$	0.03450	$3.8 \cdot 10^{-7}$	0.03063	$4.4 \cdot 10^{-7}$	0.01495
10^{-4}	$2.1 \cdot 10^{-10}$	0.00346	$2.6 \cdot 10^{-8}$	0.00437	$5.1 \cdot 10^{-8}$	0.00176

Table 1.1: Best approximation errors in $L^q(0,1)$ depending on δ_{rel}

In the numerical example we perturb the exact data with a random noise: for given $\delta_{rel} \geq 0$ we set

$$y^\delta := y + \frac{e}{\|e\|}\|y\|\,\delta_{rel}, \qquad e := \sum_{j=1}^{n} e_j \varphi_j(t), \quad t \in [0,1],$$

where e is a piece-wise constant function with Gaussian variables $e_j \sim N(0,1)$, $1 \leq j \leq n$. Here δ_{rel} denotes the relative size of noise. Since we are interested in the reconstruction potential of the underlying regularization approaches we calculate regularized solutions x_α^δ for a number of regularization parameters $a_1, \ldots, \alpha_l > 0$, $l \in \mathbb{N}$, and choose $\alpha = \alpha_{opt}$ such that $x_{\alpha_{opt}}^\delta$ gives the best approximation of $x^\dagger$ (in the norm of the space $\mathcal{X}$). Therefore we have to specify now the solution space $\mathcal{X}$. Two approaches are possible. On the one hand the specific penalty term naturally suggests to choose the Banach space $\mathcal{X} = L^q(0,1)$. This variant has the disadvantage that we also measure the regularization error $\|x_\alpha^\delta - x^\dagger\|_{L^q}$ in different norms for different choices of q. So we cannot compare the quality of the optimal regularized solutions depending on the parameter q. Therefore it is reasonable also to compare the reconstruction chances for the different choices of q by searching the best approximation in $\mathcal{X} = L^2(0,1)$.

In our experiment we consider three different choices for q, i.e. we set $q = 2$, $q = 1.5$ and $q = 1.1$. Then for different noise levels δ_{rel} we calculated the best approximations $x_{\alpha_{op}}^\delta$ in $L^q(0,1)$ and $L^2(0,1)$. The results can be found in Table 1.1 and Table 1.2. Both tables contain the relative errors of the achieved approximations as well as the corresponding regularization parameters $\alpha = \alpha_{opt}$. In particular, the results presented in Table 1.2 where all approximation errors are given with respect to the L^2-norm we observe that the choice $q = 1.1$ leads to the best recoveries for all different noise levels.

The different quality of the approximate solutions can be also observed in Figure 1.2 and 1.3. For a noise level of $\delta_{rel} = 0.01$ the two plots in Figure 1.2 contain the achieved regularized solutions with the classical Hilbert space setting $q = 2$. In the left plot (best approximation) the peaks of the exact solution are well recovered but the regularized solution oscillates at the interval where the exact solution vanishes. An increase of the regularization parameter α (see right plot) smooths out these oscillations but leads in parallel to a worse recovery of the heights of the peaks.

Figure 1.3 shows the best approximations for the same noisy data for the choices $q = 1.5$ and $q = 1.1$ measuring the regularization error in $\mathcal{X} = L^q(0,1)$. In particular, for $q = 1.1$

	$q=2$		$q=1.5$		$q=1.1$	
δ_{rel}	α_{opt}	$\frac{\|x_\alpha^\delta - x^\dagger\|_2}{\|x^\dagger\|_2}$	α_{opt}	$\frac{\|x_\alpha^\delta - x^\dagger\|_2}{\|x^\dagger\|_2}$	α_{opt}	$\frac{\|x_\alpha^\delta - x^\dagger\|_2}{\|x^\dagger\|_2}$
10^{-1}	$4.1 \cdot 10^{-5}$	0.39398	$6.4 \cdot 10^{-5}$	0.31872	$3.5 \cdot 10^{-4}$	0.32447
10^{-2}	$3.5 \cdot 10^{-6}$	0.19409	$3.5 \cdot 10^{-6}$	0.13867	$5.1 \cdot 10^{-6}$	0.12307
10^{-3}	$2.1 \cdot 10^{-8}$	0.03450	$2.4 \cdot 10^{-7}$	0.02594	$2.8 \cdot 10^{-7}$	0.01400
10^{-4}	$2.1 \cdot 10^{-10}$	0.00346	$1.3 \cdot 10^{-8}$	0.00339	$2.6 \cdot 10^{-8}$	0.00152

Table 1.2: Best approximation errors in $L^2(0,1)$ depending on δ_{rel}

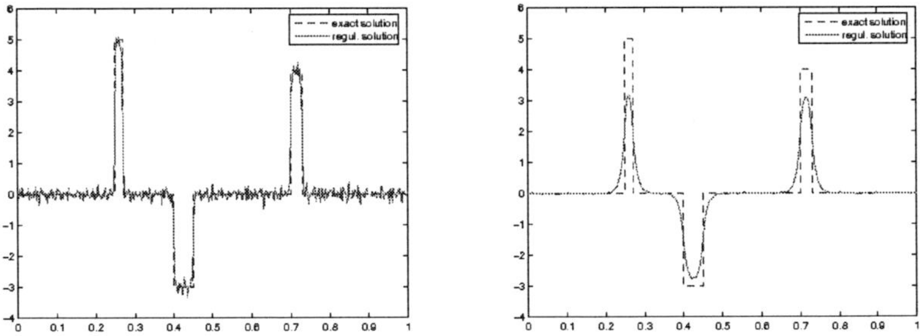

Figure 1.2: Regularized solutions for $q = 2$ with optimal α (left plot) and α chosen too large (right plot), $\delta_{rel} = 0.01$

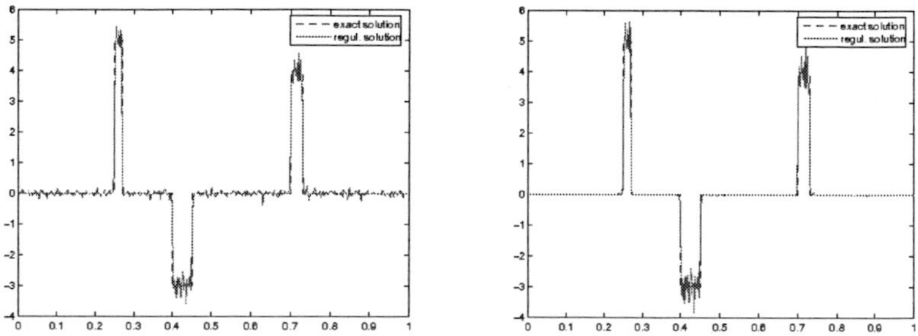

Figure 1.3: Regularized solutions for $q = 1.5$ (left plot) and $q = 1.1$ (right plot) with optimal α, $\delta_{rel} = 0.01$

(right plot) the oscillations nearly vanish on the zero intervals of the exact solution $x^\dagger$. On the other hand – on the non-zero part – the oscillations are largest for $q = 1.1$ which leads to the speculation that the choice $q = 1.1$ might be disadvantageous for non-vanishing functions (having large values). This once more emphasizes the important role of the 'proper' choice of the penalty term with respect to the reconstruction chances. This choice clearly can be based only on some a priori information about the property of the expected solution $x^\dagger$. So it is reasonable to provide a general regularization theory for Tikhonov regularization approaches with covers a wide class of possible penalty terms.

Chapter 2

Preliminaries

2.1 Geometry of Banach spaces

We start with some elementary definitions dealing with the geometry of Banach spaces, which we will need later in our convergence rate theory. For an overview and further results concerning this topic we refer e.g. to [13] and [65].

Let $r, r^* \in (1, \infty)$ and $s, s^* \in (1, \infty)$ be conjugate exponents such that

$$\frac{1}{r} + \frac{1}{r^*} = 1 \quad \text{and} \quad \frac{1}{s} + \frac{1}{s^*} = 1.$$

Throughout the monograph the spaces $\mathcal{X}$ and $\mathcal{Y}$ are assumed to be real Banach spaces with their dual spaces $\mathcal{X}^*$ and $\mathcal{Y}^*$, respectively. The norms will be denoted with $\|\cdot\|$. We omit the indices indicating the space as long as it will become clear from the context which one is meant. With $\langle \cdot, \cdot \rangle$ we denote the corresponding duality products by skipping the indices again. If $\mathcal{X}$ is supposed to be a Hilbert space then we also write $\langle \cdot, \cdot \rangle_{\mathcal{X}}$ or $\langle \cdot, \cdot \rangle$ for the underlying scalar product.

The first definition describes convexity and smoothness as elementary properties of the underlying Banach space, see e.g. [96].

Definition 2.1.1 *Let $\mathcal{X}$ be a real Banach space and $\mathcal{S}_X$ denotes the unit sphere in $\mathcal{X}$.*

(*i*) *The function $\delta_X : [0, 2] \longrightarrow [0, 1]$ defined by*

$$\delta_X(\varepsilon) := \inf \left\{ 1 - \frac{1}{2}\|x + \tilde{x}\| \ : \ x, \tilde{x} \in \mathcal{S}_X, \ \|x - \tilde{x}\| \geq \varepsilon \right\}$$

is called the modulus of convexity of $\mathcal{X}$.

(*ii*) *The function $\rho_X : [0, \infty) \longrightarrow [0, \infty)$ defined by*

$$\rho_X(\tau) := \frac{1}{2} \sup \left\{ \|x + \tilde{x}\| + \|x - \tilde{x}\| - 2 \ : \ x \in \mathcal{S}_X, \ \|\tilde{x}\| \leq \tau \right\}$$

is referred to as the modulus of smoothness of $\mathcal{X}$.

In particular, we are interested in the geometric properties of L^p-spaces. Therefore, we give the following example, see e.g. [30], [65] and [96].

Example 2.1.1 *For the spaces l^p, L^p and W^p_m, $1 < p < \infty$, $m \in \mathbb{N}$, we have for the modulus of convexity*

$$\delta_X(\varepsilon) = \begin{cases} \frac{p-1}{8}\varepsilon^2 + o(\varepsilon^2) > \frac{p-1}{8}\varepsilon^2, & 1 < p < 2, \\ 1 - \left[1 - \left(\frac{\varepsilon}{2}\right)^p\right]^{\frac{1}{p}} > \frac{1}{p}\left(\frac{\varepsilon}{2}\right)^p, & p \geq 2, \end{cases}$$

and for the modulus of continuity

$$\rho_X(\varepsilon) = \begin{cases} (1+\tau^p)^{\frac{1}{p}} - 1 < \frac{1}{p}\tau^p, & 1 < p < 2, \\ \frac{p-1}{2}\tau^2 + o(\tau^2) < \frac{p-1}{2}\tau^2, & p \geq 2. \end{cases}$$

Furthermore we introduce some definitions related to the smoothness and convexity of the underlying space $\mathcal{X}$.

Definition 2.1.2 *The Banach space $\mathcal{X}$ is called*

(*i*) *uniformly convex, if $\delta_X(\varepsilon) > 0$ for all $\varepsilon > 0$,*

(*ii*) *strictly convex, if $\|\lambda x + (1-\lambda)y\| < 1$ for all $\lambda \in (0,1)$ and all $x, y \in \mathcal{S}_X$ with $x \neq y$,*

(*iii*) *convex of power type or s-convex if for some $r > 1$ and constant $C_X > 0$*

$$\delta_X(\varepsilon) \geq C_X \varepsilon^s, \qquad \varepsilon \in [0,2],$$

(*iv*) *smooth, if for every $x \neq 0$, there exists a unique $x^* \in \mathcal{X}^*$ such that $\|x^*\| = 1$ and $\langle x^*, x\rangle = \|x\|$,*

(*v*) *uniformly smooth if $\rho_X(\tau)/\tau \to 0$ for $\tau \to 0$ and*

(*vi*) *smooth of power type or r-smooth if for some $r > 1$ and constant $\hat{C}_X > 0$*

$$\rho_X(\tau) \leq \hat{C}_X \tau^r, \qquad \tau \in [0,\infty).$$

We summerize the following properties of the moduli of convexity and smoothness, see e.g [65] and [96].

Proposition 2.1.1 *Let $\mathcal{X}$ denotes a real Banach space with modulus of convexity $\delta_X(\varepsilon)$ and modulus of smoothness $\rho_X(\tau)$. Then the following properties hold true:*

(*i*) *$\delta_X(0) = 0$, $\delta_X(\varepsilon) \leq 1$, $\delta_X(\varepsilon) < 1$ if $\mathcal{X}$ is uniformly convex.*

(*ii*) *$\delta_X(\varepsilon)$ is continuous and nondecreasing, $\delta_X(\varepsilon)$ is strictly increasing if and only if $\mathcal{X}$ is uniformly convex.*

(*iii*) *$\delta_X(\varepsilon) \leq \delta_H(\varepsilon) = 1 - \sqrt{1 - \varepsilon^2/4}$ where $\mathcal{H}$ denotes an arbitrary Hilbert space.*

(*iv*) *If $\mathcal{X}$ is is uniformly convex, then $\delta_X(\varepsilon)/\varepsilon$ is nondecreasing.*

(*v*) *$\rho_X(0) = 0$, $\rho_X(\tau) \leq \tau$.*

(*vi*) *$\rho_X(\tau)$ is convex, continuous and nondecreasing.*

(vii) $\rho_X(\tau)/\tau$ *is nondecreasing.*

(viii) $\rho_X(\tau) \geq \rho_H(\tau) = \sqrt{1+\tau^2} - 1.$

(ix) $\rho_{X^*}(\tau) = \sup\{\tau\varepsilon/2 - \delta_X(\varepsilon) \; : \; \varepsilon \in [0,2]\}.$

In particular, Proposition 2.1.1(ix) shows that smoothness of a space $\mathcal{X}$ is closely related to the convexity of the dual space $\mathcal{X}^*$. The duality of both concepts also shows the following result, see e.g. [13] or [88, Theorem 2.5].

Proposition 2.1.2 *Let $\mathcal{X}$ denotes a real Banach space with dual space $\mathcal{X}^*$.*

(i) *$\mathcal{X}$ is uniformly convex $\Leftrightarrow$ $\mathcal{X}^*$ is uniformly smooth.*

(ii) *If $\mathcal{X}$ is uniformly convex then $\mathcal{X}$ is reflexive and strictly convex.*

(iii) *If $\mathcal{X}$ is uniformly smooth then $\mathcal{X}$ is reflexive and smooth.*

(iv) *Let $\mathcal{X}$ be reflexive. Then $\mathcal{X}$ is strictly convex $\Leftrightarrow$ $\mathcal{X}^*$ is smooth.*

(v) *$\mathcal{X}$ is r-smooth and s-convex $\Leftrightarrow$ $\mathcal{X}^*$ is r^*-convex and s^*-smooth. Moreover, $r \in [2,\infty)$ and $s \in (1,2]$ holds.*

In particular, the fact that every uniformly convex Banach space is reflexive is well-established as Milman-Pettis Theorem, see [70] and [78]. The point (v) is an immediate of the Lindenstrauss duality formula, see [65, IV.1.7 and IV.1.12] and Dvoretzky's theorem [18].

Example 2.1.2 *The spaces l^p, L^p and W^p_m, $1 < p < \infty$, $m \in \mathbb{N}$, are uniformly convex as well as uniformly smooth, see [65]. Moreover, Example 2.1.1 above shows, that these spaces are $\max\{p,2\}$-convex and $\min\{p,2\}$-smooth. For $p = 1$ or $p = \infty$ these spaces are neither smooth nor strictly convex. Clearly, if $\mathcal{X}$ is a Hilbert space then the polarization identity*

$$\|x + \tilde{x}\|^2 = \|x\|^2 + 2\langle x, \tilde{x}\rangle_{\mathcal{X}} + \|\tilde{x}\|^2, \quad \forall x, \tilde{x} \in \mathcal{X},$$

shows that any Hilbert space is 2-convex and 2-smooth.

2.2 Duality mappings

We define the duality mappings, see e.g. [13], [98, Section 47.12] and [96] which play a key role in the subsequent analysis.

Definition 2.2.1 *Let $q > 1$. The set-valued mapping $J_q : \mathcal{X} \longrightarrow 2^{\mathcal{X}^*}$ which is defined as*

$$J_q(x) := \left\{x^* \in \mathcal{X}^* \; : \; \langle x^*, x\rangle = \|x^*\|\,\|x\|,\, \|x^*\| = \|x\|^{q-1}\right\}$$

is called the duality mapping of $\mathcal{X}$ with gauge function $t \mapsto t^{q-1}$. The duality mapping $J = J_2$ is also referred to as the normalized duality mapping.

With $2^{\mathcal{X}^*}$ we denote the power set of the space $\mathcal{X}^*$. The properties of duality mappings are well-studied, see e.g. [13, Chapter 1 and 2] and [98, Proposition 47.17]. Studying Tikhonov regularization approaches these properties play an important role in the convergence rates analysis as well as in the development of numerical algorithms for solving the underlying minimization problems. Therefore we summarize the following results.

Proposition 2.2.1 *Let $\mathcal{X}$ denotes a Banach space with duality mapping J_q, $q > 1$. Then we have the following properties:*

(i) If $\mathcal{X}$ is a Hilbert space, then $J_2 =: J$ (respectively J_2^{-1}) is the Riesz isomorphism.

(ii) For all $\lambda \in \mathbb{R}$ and all $x \in \mathcal{X}$ we have

$$J_q(\lambda x) = sgn(\lambda)|\lambda|^{q-1}\|x\|^{q-2}J(x).$$

(iii) J_q is monotone and J_q is strictly monoton $\Leftrightarrow$ X is strictly convex.

(iv) $J_q(x) = \partial\left\{\frac{1}{q}\|x\|^q\right\}$ where $\partial\{\cdot\}$ denotes the subdifferential.

(v) J_q is always single valued $\Leftrightarrow$ X is smooth.

(vi) Let X be reflexive, strictly convex and smooth. Then J_q is norm-to-weak-continuous and bijective and $J_q^{-1} = J^*_{q^*} : \mathcal{X}^* \longrightarrow \mathcal{X}$ where $J^*_{q^*}$ is the duality mapping in $\mathcal{X}^*$ with gauge function $t \mapsto t^{q^*-1}$.

(vii) If $\mathcal{X}$ and $\mathcal{X}^*$ are uniformly convex then J_q and J_q^{-1} are (norm-)continuous on $\mathcal{X}$ and $\mathcal{X}^*$, respectively.

Here, q^* is the conjugate exponent of q, i.e. $q^{-1} + (q^*)^{-1} = 1$. In this context we recapitulate that we call a mapping $F : \mathcal{X} \longrightarrow \mathcal{Y}$

- monotone, if $\mathcal{Y} = \mathcal{X}^*$ and

$$\langle F(x) - F(y), x - y\rangle \geq 0, \qquad \forall\, x, y \in \mathcal{X},$$

- strictly monotone, if $\mathcal{Y} = \mathcal{X}^*$ and

$$\langle F(x) - F(y), x - y\rangle > 0, \qquad \forall\, x, y \in \mathcal{X},$$

  as long as $x \neq y$ and
- norm-to-weak-continuous if norm convergence $x_n \longrightarrow x$ in $\mathcal{X}$ implies weak convergence $F(x_n) \rightharpoonup F(x)$ in $\mathcal{X}^*$.

In particular, the point (iv) (known also as Theorem of Asplund, see [13, Theorem I.4.4]) takes a key role in our further considerations. Moreover, if not specified otherwise we will always suppose smooth Banach spaces. Then we can presume the duality mappings to be single valued. This assumption in particular excludes the spaces L^1 and L^∞ in accordant considerations.

The Xu/Roach-inequalities, see [96], connect duality mappings with the smoothness and convexity of the corresponding Banach space.

Proposition 2.2.2 (Xu/Roach inequalities) *Let $\mathcal{X}$ be a real Banach space with modulus of convexity δ_X and modulus of smoothness ρ_X and $1 < r, s < \infty$. For given $\tilde{x}, x \in \mathcal{X}$ we set $H(t) := \max\{\|x + t\,\tilde{x}\|, \|x\|\}$.*

(i) *If $\mathcal{X}$ is uniformly convex with duality mapping J_s then the inequality*

$$\frac{1}{s}\|x+\tilde{x}\|^s \geq \frac{1}{s}\|x\|^s + \langle x^*, \tilde{x}\rangle + K_s \int_0^1 \frac{H(t)^s}{t}\delta_X\left(\frac{t\|\tilde{x}\|}{2\,H(t)}\right) dt$$

holds for all $x, \tilde{x} \in \mathcal{X}$ where

$$\begin{aligned} K_s \;&:=\; 4\left(2+\sqrt{3}\right)\min\left\{\min\left\{\frac{1}{2}s(s-1), 1\right\}, \min\left\{\frac{1}{2}s, 1\right\}(s-1),\right. \\ &\qquad\left.(s-1)\left(1-\left(\sqrt{3}-1\right)^{s^*}\right), 1-\left(1+\left(2-\sqrt{3}\right)s^*\right)^{1-s}\right\} \end{aligned}$$

and $x^* \in J_s(x)$ is arbitrary chosen.

(ii) *If $\mathcal{X}$ is uniformly smooth with duality mapping J_r then the inequality*

$$\frac{1}{r}\|x+\tilde{x}\|^r \leq \frac{1}{r}\|x\|^r + \langle J_r(x), \tilde{x}\rangle + L_r \int_0^1 \frac{H(t)^r}{t}\rho_X\left(\frac{t\|\tilde{x}\|}{H(t)}\right) dt$$

holds for all $x, \tilde{x} \in \mathcal{X}$ where $L_r := \max\left\{8, 64\,c\,K_{r^*}^{-1}\right\}$ with

$$c := \frac{4\tau_0}{\rho_H(\tau_0)}\prod_{k=0}^{\infty}\left(1+\frac{15\tau_0}{4\,2^k}\right) \quad \text{with} \quad \tau_0 := \frac{\sqrt{339}-18}{30}.$$

We are in particular interested in the situation that r and s describe the modulus of convexity and smoothness respectively. Let therefore $\mathcal{X}$ supposed to be a s-convex Banach space, i.e. $\delta_X(\varepsilon) \geq C_X\varepsilon^s$, $\varepsilon \in [0,2]$. Then the integral simplifies to

$$K_s C_X \int_0^1 \frac{H(t)^s}{t}\left(\frac{t\|\tilde{x}\|}{2\,H(t)}\right)^s dt = \frac{K_s C_X}{2^s s}\|\tilde{x}\|^s.$$

On the other hand, for r-smooth spaces we have $\rho_X(\tau) \leq \hat{C}_X\tau^r$, $\tau \in [0,\infty)$ for some constant $\hat{C}_X > 0$ and hence

$$L_r\hat{C}_X \int_0^1 \frac{H(t)^r}{t}\left(\frac{t\|\tilde{x}\|}{H(t)}\right)^r dt = \frac{L_r\hat{C}_X}{r}\|\tilde{x}\|^r.$$

Consequently, with $G_r := L_r\hat{C}_X$ and $C_s := K_s C_X 2^{-s}$ the following holds.

Corollary 2.2.1 *Let $\mathcal{X}$ be a real Banach space.*

(i) *If $\mathcal{X}$ is s-convex, then there exists a constant $C_s > 0$ such that*

$$\frac{1}{s}\|x+\tilde{x}\|^s \geq \frac{1}{s}\|x\|^s + \langle x^*, \tilde{x}\rangle + \frac{C_s}{s}\|\tilde{x}\|^s, \qquad \forall\, x, \tilde{x} \in \mathcal{X},$$

where $x^* \in J_s(x)$ is arbitrary chosen.

(ii) *If $\mathcal{X}$ is r-smooth, then there exists a constant $G_r > 0$ such that*

$$\frac{1}{r}\|x+\tilde{x}\|^r \leq \frac{1}{r}\|x\|^p + \langle J_r(x), \tilde{x}\rangle + \frac{G_r}{r}\|\tilde{x}\|^r, \qquad \forall\, x, \tilde{x} \in \mathcal{X}.$$

We also remark, that the constant L_r comes from the estimate

$$\|J_r(\tilde{x}+x) - J_r(x)\| \leq K_{r^*}\frac{\max\{\|\tilde{x}+x\|, \|x\|\}^r}{\|\tilde{x}\|}\rho_X\left(\frac{8\|\tilde{x}\|}{K_{r^*}\max\{\|\tilde{x}+x\|, \|x\|\}}\right).$$

Then, for the r-smooth space $\mathcal{X}$ we have

$$\|J_r(\tilde{x}+x) - J_r(x)\| \leq K_{r^*}^{1-p}\hat{C}_X 8^r \|\tilde{x}\|^{r-1},$$

i.e. we can choose $G_r := K_{r^*}^{1-r}\hat{C}_X 8^r$.

We discuss duality mapping in L^p-spaces.

Example 2.2.1 *For simplicity, we assume $\mathcal{X} = L^p(0,1)$, $1 < p < \infty$, and $q > 0$. We introduce the notation $\|\cdot\|_p := \|\cdot\|_{L^p(0,1)}$. We define the functional*

$$P(x) := \frac{1}{q}\|x\|_p^q = \frac{1}{q}\left(\int_0^1 |x(t)|^p\, dt\right)^{\frac{q}{p}}, \qquad x \in L^p(0,1).$$

Then for $x, h \in \mathcal{X}$ we derive

$$\begin{aligned}\lim_{\varepsilon\to 0}\frac{P(x+\varepsilon\, h) - P(x)}{\varepsilon} &= \lim_{\varepsilon\to 0}\frac{1}{q\varepsilon}\left(\left(\int_0^1 |(x+\varepsilon\, h)(t)|^p\, dt\right)^{\frac{q}{p}} - \left(\int_0^1 |x(t)|^p\, dt\right)^{\frac{q}{p}}\right)\\ &= \|x\|_p^{q-p}\int_0^1 |x(t)|^{p-1} sgn(x(t))h(t)\, dt,\end{aligned}$$

i.e. $J_q(x) = \|x\|_p^{q-p}|x|^{p-1}sgn(x(\cdot))$. Moreover, with $p^* := \frac{p}{p-1}$ we have

$$\|J_q(x)\|_{p^*} = \|x\|_p^{q-p}\left(\int_0^1 |x|^{(p-1)p^*}\, dt\right)^{\frac{1}{p^*}} = \|x\|_p^{q-p}\|x\|_p^{\frac{p-1}{p}p} = \|x\|_p^{q-1}$$

which shows $J_q(x) \in L^p(0,1)^* = L^{p^*}(0,1)$ and $\|J_q(x)\|_{p^*} = \|x\|_p^{q-1}$ which was part of the definition of J_q.

We discuss differentiability of the duality mapping. Therefore we have to assume $p \geq 2$ and $x \neq 0$ for $q < 2p$. For $p > 2$ and $J_q'(x) : \mathcal{X} \longrightarrow \mathcal{X}^*$ we calculate

$$J_q'(x) = \left(\|x\|_p^{q-p}\right)' |x|^{p-1} sgn(x(\cdot)) + \|x\|_p^{q-p}(p-1)|x|^{p-2}$$

whereas, for $p = 2$ we obtain

$$J_q'(x) = \left(\|x\|_2^{q-2}\right)' x + \|x\|_2^{q-2}$$

by using $|x| sgn(x(\cdot)) = x$. We introduce the function

$$m := \begin{cases} \|x\|_p^{q-p}(p-1)|x|^{p-2}, & p > 2, \\ \|x\|_2^{q-2}, & p = 2. \end{cases}$$

Furthermore, we have

$$\left(\|x\|_p^{q-p}\right)' = \left[\left(\|x\|_p^q\right)^{\frac{q-p}{q}}\right]' = \frac{q-p}{q}\left(\|x\|_p^q\right)^{\frac{q-p}{q}-1} q\, J_q(x) = (q-p)\|x\|_p^{-p} J_q(x).$$

Hence, for arbitrary $h_1, h_2 \in L^p(0,1)$ we obtained

$$\begin{aligned} J_q'(x)(h_1,h_2) &= (q-p)\|x\|_p^{q-2p} \int_0^1 |x(t)|^{p-1} sgn(x(t)) h_1(t)\, dt \int_0^1 |x(t)|^{p-1} sgn(x(t)) h_2(t)\, dt \\ &\quad + \int_0^1 m(t)\, h_1(t)\, h_2(t)\, dt \\ &=: M_1(h_1,h_2) + M_2(h_1,h_2). \end{aligned}$$

Since $\||x|^{p-1}\|_{p^*} = \|x\|_p^{p-1}$ we then conclude by Hölder's inequality

$$\begin{aligned} |M_1(h_1,h_2)| &\leq |q-p|\|x\|_p^{q-2p}\||x|^{p-1}\|_{p^*}\|h_2\|_p\||x|^{p-1}\|_{p^*}\|h_1\|_p \\ &= |q-p|\|x\|_p^{q-2}\|h_1\|_p\|h_2\|_p \end{aligned}$$

and

$$|M_2(h_1,h_2)| \leq \|m\|_{\hat{p}}\|h_1\|_p\|h_2\|_p$$

with $\hat{p} > 1$ chosen such that

$$\frac{1}{\hat{p}} + \frac{2}{p} = 1 \Leftrightarrow \hat{p} = \frac{p}{p-2} \quad \text{for} \quad p > 2$$

and $\hat{p} = \infty$ for $p = 2$. For $p > 2$ we get

$$\||x|^{p-2}\|_{\hat{p}} = \left(\int_0^1 |x(t)|^{(p-2)\frac{p}{p-2}}\, dt\right)^{\frac{p-2}{p}} = \|x\|_p^{p-2},$$

which shows $\|m\|_{\hat{p}} \le (p-1)\|x\|_p^{q-2}$ *for arbitrary* $p \ge 2$. *This proves*

$$\|J_q'(x)\| \le (p - 1 + |q-p|)\|x\|_p^{q-2}.$$

Hence $\langle J_q'(x)\cdot, \cdot\rangle : L^p(0,1) \times L^p(0,1) \longrightarrow \mathbb{R}$ *defines a continuous and symmetric bilinear form. In particular, for* $q = 2$, *the norm of* $J_q'(x)$ *does not depend on the specific element* $x \in L^p(0,1)$ *and we have* $\|J_q'(x)\| \le 2p - 3$. *Moreover, we are interested in* $J_q'(x)(h_1, h_1)$. *By convexity of* P *we observe* $J_q'(x)(h_1,h_1) \ge 0$ *for all* $x, h_1 \in \mathcal{X}$. *Assume* $q \le p$. *Then* $M_1(h_1,h_1) \le 0$ *holds. So this term can be dropped. For* $q > p$ *the term* $M_1(h_1,h_1)$ *is non-negative. Hence we derive*

$$|J_q'(x)(h_1,h_1)| \le \begin{cases} (p-1)\|x\|_p^{q-2}\|h_1\|^2, & p < q, \\ (q-1)\|x\|_p^{q-2}\|h_1\|^2, & p \ge q. \end{cases}$$

So we can apply a Taylor approximation for $q = 2$ *to obtain for arbitrary* $x, \tilde{x} \in L^p(0,1)$

$$\frac{1}{2}\|x+\tilde{x}\|_p^2 \le \frac{1}{2}\|x\|_p^2 + \langle J_2(x), \tilde{x}\rangle + \frac{(p-1)}{2}\|\tilde{x}\|_p^2$$

This coincides with the Xu/Roach inequalities. Since the space $L^p(0,1)$ *is 2-smooth for* $p \ge 2$, *we derived the assertion of Corollary 2.2.1(ii) with* $G_2 \le (p-1)$. *This is also the result which can be found already in [95].*

2.3 Bregman distances

In the recent years, Bregman distances of convex functionals has been well-established as powerful tool in convergence rate theory for regularization approaches in Banach spaces, see e.g. [12] and [45]. There, the definitions of Bregman distances slightly differ with respect to their set-/single-valued definition. Here we use the definition of [45] where the Bregman distance is related to one fixed element of the subdifferential. Therefore, the definition we will use in our further analysis is given as follows.

Definition 2.3.1 *Let* $P : \mathcal{D}(P) \subseteq \mathcal{X} \longrightarrow \mathbb{R}$ *denote a convex functional with domain* $\mathcal{D}(P)$. *For given* $\tilde{x}, x \in \mathcal{D}(P)$ *and* $\xi \in \partial P(x)$ *the Bregman distance is defined as*

$$D_\xi(\tilde{x}, x) := P(\tilde{x}) - P(x) - \langle \xi, \tilde{x} - x\rangle.$$

The set $\mathcal{D}_B := \{x \in \mathcal{X} : \partial P(x) \ne \emptyset\}$ *is called the Bregman domain.*

Originally, Bregman distances were introduced in [11] for strictly convex and Fréchet differentiable functionals, i.e. the subdifferential $\partial P(x) = \{P'(x)\}$ was supposed to be single valued for each $x \in \mathcal{D}(P)$. A further generalization to non-smooth functionals which are strictly convex is given in [57]. Even more generally, since the penalty functional P in context with regularization might be neither smooth nor strictly convex, Bregman distances were introduced in [12] as family $D_\xi(\tilde{x}, x) := \{P(\tilde{x}) - P(x) - \langle \xi, \tilde{x} - x\rangle : \xi \in \partial P(x)\}$ for all $\tilde{x}, x \in \mathcal{D}(P)$.

Obviously, we have $D_\xi(\tilde{x},x) \geq 0$ for all $\tilde{x}, x \in \mathcal{D}(P)$. The relation $D_\xi(\tilde{x},x) = 0$ implies $\tilde{x} = x$ if and only if P is strictly convex. Moreover, even for strictly convex functionals, Bregman distances in general do not define metrics since they are neither symmetrically nor transitive in general.

We shortly present the following standard example.

Example 2.3.1 *Assume $\mathcal{X}$ to be smooth and $1 < q < \infty$. Then we define $P : \mathcal{X} \longrightarrow [0,\infty)$ as $P(x) := \frac{1}{q}\|x\|^q$, $x \in \mathcal{X}$. Let $x, \tilde{x} \in \mathcal{X}$ chosen arbitrary. From the properties of the duality mapping we have $\xi := P'(x) = J_q(x)$ and*

$$\begin{aligned} D_\xi(\tilde{x},x) &:= \frac{1}{q}\|\tilde{x}\|^q - \frac{1}{q}\|x\|^q - \langle J_q(x), \tilde{x} - x\rangle \\ &= \frac{1}{q}\|\tilde{x}\|^q + \frac{1}{q^*}\|x\|^q - \langle J_q(x), \tilde{x}\rangle \\ &= \frac{1}{q}\|\tilde{x}\|^q + \frac{1}{q^*}\|J_q(x)\|^{q^*} - \langle J_q(x), \tilde{x}\rangle, \end{aligned}$$

by using $\langle J_q(x), x\rangle = \|x\|^q$ and $\|J_q(x)\| = \|x\|^{q-1}$. Furthermore, let $\mathcal{X}$ be reflexive and we assume the dual space $\mathcal{X}^*$ to be smooth as well. We introduce the functional $P_* : \mathcal{X}^* \longrightarrow [0,\infty)$ as $P_*(x^*) := \frac{1}{q^*}\|x^*\|^{q^*}$, $x^* \in \mathcal{X}^*$. With $\tilde{x} := P_*'(\tilde{x}^*) = J_{q^*}(\tilde{x}^*) = J_q^{-1}(\tilde{x}^*)$ we conclude

$$\begin{aligned} D_x^*(J_q(x), \tilde{x}^*) &:= \frac{1}{q^*}\|J_q(x)\|^{q^*} - \frac{1}{q^*}\|\tilde{x}^*\|^{q^*} - \langle \tilde{x}, J_q(x) - \tilde{x}^*\rangle \\ &= \frac{1}{q^*}\|J_q(x)\|^{q^*} + \frac{1}{q}\|\tilde{x}\|^q - \langle \tilde{x}, J_q(x)\rangle = D_\xi(\tilde{x},x) \end{aligned}$$

which shows an interesting correlation between the Bregman distances of the functionals P and P_*.

In the further analysis we need some conditions which provides bounds for the Bregman distances $D_\xi(\tilde{x},x)$ by the norm $\|\tilde{x} - x\|$ as $\tilde{x} \to x$. Here we can derive such estimates directly from the Xu/Roach inequalities [96]. We will show this by a short calculation. Assume therefore $\tilde{x} \in \mathcal{B}_\varrho(x)$ for some $\varrho > 0$. Then, for $t \in [0,1]$ we estimate

$$\max\{\|x + t(\tilde{x} - x)\|, \|x\|\} \leq \max\{\|x\| + t\|\tilde{x} - x\|, \|x\|\} \leq \|x\| + \varrho$$

and

$$\max\{\|x + t(\tilde{x} - x)\|, \|x\|\} \geq \|x\|.$$

Assume $\mathcal{X}$ to be r-smooth, i.e. have the modulus of smoothness $\rho_X(\tau) \leq \hat{C}_X \tau^r$, $\tau \in [0,\infty)$ for some constant $\hat{C}_X > 0$. Then, from Proposition 2.2.2(ii) we conclude

$$D_\xi(\tilde{x},x) \leq L_q \hat{C}_X \int_0^1 \max\{\|x + t(\tilde{x} - x)\|, \|x\|\}^{q-r} t^{r-1}\, dt\, \|\tilde{x} - x\|^r \leq \frac{G_r(x)}{r}\|\tilde{x} - x\|^r$$

with

$$G_r(x) := \begin{cases} L_q \hat{C}_X (\|x\| + \varrho)^{q-r}, & q > r, \\ L_q \hat{C}_X, & q = r, \\ L_q \hat{C}_X \|x\|^{q-r}, & q < r. \end{cases}$$

Assume now that $\mathcal{X}$ is s-convex for some $1 < s \le 2$. Then a similar calculation leads to a lower bound for the Bregman distance $D_\xi(\tilde{x}, x)$, i.e. we can find a constant $C_s(x) > 0$ such that

$$D_\xi(\tilde{x}, x) \ge \frac{C_s(x)}{s} \|\tilde{x} - x\|^s$$

holds for all $\tilde{x} \in \mathcal{B}_\varrho(x)$. In particular, if $\mathcal{X}$ is an L^p-space, $1 < p < \infty$, then we can immediately choose $s := \min\{2, p\}$ and $r := \max\{2, p\}$.

2.4 General vs. approximate source conditions

In order to verify convergence rates for linear (and nonlinear) regularization approaches in Hilbert spaces two concepts have been established in the recent years. We briefly summarize both ideas. Introducing further notations we recall the definition of a linear regularization $\{g_\alpha\}$, $0 < \alpha \le \|A\|^2$, for the linear problem (1.2), see e.g. [69] and [21, Chapter 3]. Therefore we assume here $\mathcal{X}$ and $\mathcal{Y}$ to be Hilbert spaces.

Definition 2.4.1 *A family $\{g_\alpha\}$, $0 < \alpha \le \|A\|^2$, of piecewise continuous functions is called a regularization if there are constants $C_1 > 0$ and $C_2 > 0$ such that for $0 < \alpha \le \|A\|^2$*

$$\sup_{0 < t \le \|A\|^2} \sqrt{t}|g_\alpha(t)| \le \frac{C_1}{\sqrt{\alpha}} \quad \text{and} \quad \sup_{0 < t \le \|A\|^2} |1 - t\, g_\alpha(t)| \le C_2. \tag{2.1}$$

For given noisy data y^δ and regularization parameter $\alpha > 0$ then the regularized approximate solution x_α^δ of (1.2) is defined as

$$x_\alpha^\delta := g_\alpha(A^*A)A^*y^\delta.$$

An important role in the analysis of general source conditions play the so-called index functions. We briefly recall the definition, see e.g. [31] and [46].

Definition 2.4.2 *A real function $\varphi(t)$, $t > 0$, is called an index function if it is continuous and strictly increasing with $\varphi(t) \to 0$ as $t \to 0$.*

For obtaining estimates of the regularization error we need the concept of (general) qualifications, see [69, Definition 1]. We also refer to [91] where this topic was introduced by a somewhat different notation.

Definition 2.4.3 *The regularization $\{g_\alpha\}$ is said to have qualification $\varphi(t)$, $t \ge 0$, for an index function $\varphi(t)$, if there exists a constant $C_\varphi > 0$ such that*

$$\sup_{0 < t \le \|A\|^2} |1 - t\, g_\alpha(t)|\varphi(t) \le C_\varphi \varphi(\alpha), \quad 0 < \alpha \le \|A\|^2.$$

First we deal with general source conditions. Let $\mathcal{R}(A)$ and $\mathcal{N}(A)$ denote the range and the kernel of the operator. Further we assume $y \in \mathcal{D}(A^\dagger) := \mathcal{R}(A) \oplus \mathcal{R}(A)^\perp$, where $A^\dagger$ describes the Moore-Penrose inverse of A, see e.g. [21, Section 2.1] and [73]. For given index function $\varphi(t)$, $t \geq 0$, we assume that $x^\dagger = A^\dagger y$ satisfies the source condition

$$x^\dagger = \varphi(A^*A)\,\omega, \qquad \omega \in \mathcal{X},\ \|\omega\| \leq R, \tag{2.2}$$

for some $R > 0$. Then, if this function $\varphi(t)$, $t \geq 0$, is a qualification of the regularization $\{g_\alpha\}$, we obtain the error estimate

$$\|x_\alpha^\delta - x^\dagger\| \leq \frac{C_1}{\sqrt{\alpha}}\delta + C_\varphi \varphi(\alpha) R, \qquad 0 < \alpha \leq \|A\|^2, \tag{2.3}$$

see [69]. Finally, an a-priori parameter choice $\alpha := \Psi^{-1}(\delta)$ with $\Psi(\alpha) := \sqrt{\alpha}\varphi(\alpha)$ leads to the convergence rate

$$\|x_\alpha^\delta - x^\dagger\| = \mathcal{O}\left(\varphi\left(\Psi^{-1}(\delta)\right)\right) \quad \text{as} \quad \delta \to 0. \tag{2.4}$$

Alternatively we can assume, that the source condition (2.2) with given index function $\varphi(t)$, $t \geq 0$, is violated. Therefore we introduce the sets

$$\mathcal{M}_\varphi(R,d) := \{x \in \mathcal{X} \,:\, x = \varphi(A^*A)\omega + v,\ \|\omega\| \leq R,\ \|v\| \leq d\}\,. \tag{2.5}$$

for parameters $R, d \geq 0$. Then, for $x^\dagger \in \mathcal{M}_\varphi(R,d)$, i.e. $x^\dagger = \varphi(A^*A)\omega + v$ with $\omega, v \in \mathcal{X}$ satisfying $\|\omega\| \leq R$ and $\|v\| \leq d$, we conclude from classical linear regularization theory

$$\begin{aligned}
\|x_\alpha^\delta - x^\dagger\| &= \|g_\alpha(A^*A)A^*(y^\delta - y + y) - x^\dagger\| \\
&\leq \|(I - g_\alpha(A^*A)A^*A)x^\dagger\| + \|g_\alpha(A^*A)A^*(y^\delta - y)\| \\
&= \|(I - g_\alpha(A^*A)A^*A)\left[\varphi(A^*A)\omega + v\right]\| + \|g_\alpha(A^*A)A^*(y - y^\delta)\| \\
&\leq C_\varphi \varphi(\alpha) R + C_2 d + \frac{C_1}{\sqrt{\alpha}}\delta.
\end{aligned}$$

The idea of approximate source conditions was originally introduced in [7, Theorem 6.8] for measuring the approximation term $\|x - A^*\tilde{\omega}\|$ for given $x \notin \mathcal{R}(A^*)$ with $\tilde{\omega} \in \mathcal{Y}$, $\|\tilde{\omega}\| \leq R$, for each $R > 0$. In order to prove convergence rates for approximate source conditions we additionally need the concept of distance functions. This approach was originally introduced in [42] for $\varphi(t) = t^{\frac{1}{2}}$ and Tikhonov regularization. Generalizing the idea, we present the following definition.

Definition 2.4.4 *For given $x \in \mathcal{X}$ and index function $\varphi(t)$, $t \geq 0$, the distance function $d_\varphi(\cdot; x) : [0,\infty) \longrightarrow \mathbb{R}$ (with respect to the reference source condition (2.2)) is defined as*

$$d_\varphi(R; x) := \min\left\{\|x - \varphi(A^*A)\omega\| \,:\, \omega \in \mathcal{X},\ \|\omega\| \leq R\right\}, \quad R \geq 0. \tag{2.6}$$

Note, that the nonnegative function $d_\varphi(R;x)$ is well-defined for each $x \in \mathcal{X}$. In particular, for each $R \geq 0$ there exists an element $\omega = \omega(R)$ with $d_\varphi(R;x) = \|x - \varphi(A^*A)\omega\|$ and

$\|\omega\| \leq R$. The distance functions are non-increasing with $d_\varphi(R;x) \to 0$ for $R \to \infty$ if $x \in \overline{\mathcal{R}(\varphi(A^*A))} = \mathcal{R}(A^\dagger)$ since

$$\overline{\mathcal{R}(\varphi(A^*A))} = \overline{\mathcal{R}((A^*A)^{\frac{1}{2}})} = \overline{\mathcal{R}(A^*)} = \mathcal{N}(A)^\perp = \mathcal{R}(A^\dagger).$$

We have $d_\varphi(R;x) > 0$ for all $R \geq 0$ if $x \notin \mathcal{R}(\varphi(A^*A))$ and $d_\varphi(R;x) = 0$ for all $R \geq \|\omega\|$ if $x = \varphi(A^*A)\omega$ and $\|\omega\| = R$.

With the aid of the distance function, the error bound for approximate source conditions has the same structure as in the case of general source conditions. A similar result is presented in [17, Theorem 2.5] for Tikhonov regularization with a slightly modified notation.

Theorem 2.4.1 *Let the index function $\varphi(t)$, $t \geq 0$, be a qualification of the regularization $\{g_\alpha\}$. We assume $x^\dagger \notin \mathcal{R}(\varphi(A^*A))$ has distance function $d(R) := d_\varphi(R;x^\dagger)$. Let $\Theta(R) := \varphi^{-1}(d(R)R^{-1})$. Then the estimate*

$$\|x_\alpha^\delta - x^\dagger\| \leq \frac{C_1}{\sqrt{\alpha}}\delta + (C_\varphi + C_2)\, d\left(\Theta^{-1}(\alpha)\right) \tag{2.7}$$

holds for all $\delta > 0$ and all $\alpha > 0$. Moreover, we define the functions $\Psi(\alpha) := \sqrt{\alpha} d(\Theta^{-1}(\alpha))$ and $\Phi(R) := \sqrt{\Theta^{-1}(d(R)R^{-1})}d(R)$. Then, an a-priori parameter choice $\alpha := \Psi^{-1}(\delta)$ leads to a convergence rate

$$\|x_\alpha^\delta - x^\dagger\| = \mathcal{O}\left(d\left(\Phi^{-1}(\delta)\right)\right) \quad \text{as} \quad \delta \to 0. \tag{2.8}$$

Proof. First we already noticed that $x^\dagger \in \overline{\mathcal{R}(\varphi(A^*A))}$. By definition, $x^\dagger \in \mathcal{M}_\varphi(R, d(R))$ for all $R \geq 0$. Hence, the estimate

$$\|x_\alpha^\delta - x^\dagger\| \leq \frac{C_1}{\sqrt{\alpha}}\delta + C_\varphi \varphi(\alpha) R + C_2 d(R)$$

holds for all $R \geq 0$. We choose $R = R(\alpha) > 0$ such that

$$R\,\varphi(\alpha) = d(R) \;\Leftrightarrow\; \varphi(\alpha) = \frac{d(R)}{R} \;\Leftrightarrow\; \alpha = \varphi^{-1}\left(d(R)R^{-1}\right) = \Theta(\alpha).$$

This proves (2.7). The choice $\alpha = \Psi^{-1}(\delta)$ implies

$$\frac{\delta}{\sqrt{\alpha}} = d\left(\Theta^{-1}(\alpha)\right) \;\Leftrightarrow\; \delta = \sqrt{\alpha} d\left(\Theta^{-1}(\alpha)\right).$$

On the other hand, we have

$$\frac{\delta}{\sqrt{\alpha}} = d(R) \;\Leftrightarrow\; \delta = d(R)\sqrt{\alpha} = \sqrt{\Theta^{-1}(d(R)R^{-1})}d(R) = \Phi(R).$$

The choice $R = \Phi^{-1}(\delta)$ leads to (2.8). ■

Since $d(R)$ is non-increasing the function $R \mapsto d(R)/R$ is strictly decreasing on $(0,\infty)$ with $d(R)/R \to \infty$ for $R \to 0$ and $d(R)/R \to 0$ for $R \to \infty$. This implies, that $\Theta(R)$ is strictly decreasing. Hence, $\Theta^{-1}(\alpha)$ is well-defined and strictly decreasing. Moreover, we observe that $d(\Theta^{-1}(\alpha))$ is an index function, i.e. it is increasing with $d(\Theta^{-1}(\alpha)) \to 0$ for $\alpha \to 0$. Analogous calculations show that the functions $\Psi^{-1}(\delta)$ and $\Phi^{-1}(\delta)$ are well-defined.

Remark 2.4.1 *We also want to point out that the supposed reference source condition (2.2) plays the role of a limit situation. Using the technique of distance functions we can only prove convergence rates lower that the rate (2.4) since condition (2.2) is always assumed to be violated. So if we expect higher smoothness than $x^\dagger \in \mathcal{R}(\varphi(A^*A))$ we have to suppose a stronger reference source condition.*

Comparing both error bounds (2.3) and (2.7) we notice the similar structure. In both cases the term $C_1 \frac{\delta}{\sqrt{\alpha}}$ is based on the error in the given data. The second term is the regularization error which depends on the regularization parameter α and the (approximate) source condition, which is usually unknown.

Moreover, it is of high interest to compare the result (2.7) of Theorem 2.4.1 with the estimate (2.3). Therefore we consider the following situation. The element $x \in \mathcal{X}$ does not belong to $\mathcal{R}(\varphi(A^*A))$ but it satisfies the weaker source condition $x \in \mathcal{R}(\psi(A^*A))$ for some index function $\psi(t)$, $t \geq 0$. We will show that we can find an upper bound for the distance function $d_\varphi(R;x)$, $R \geq 0$, under this additional assumption. To do so we apply a result of [46]. There, an alternative variant of the idea distance function was used to define an accordant function. In particular, the distance function $\hat{d}_\varphi(t;x)$, $t \geq 0$, was introduced as

$$\hat{d}_\varphi(t;x) := \inf \{ \|t\,x - \varphi(A^*A)\omega\| \ : \ \omega \in \mathcal{X},\ \|\omega\| \leq 1 \}.$$

It can be shown, that this function is an index function, see [46, Lemma 5.3]. So this definition has the advantage keeping the subsequent analysis in the context of index functions. Both functions $d_\varphi(R;x)$ and $\hat{d}_\varphi(t;x)$ are closely related. To verify this we rewrite for $R > 0$

$$\begin{aligned} d_\varphi(R;x) &= \inf \{ \|x - \varphi(A^*A)\omega\| \ : \ \omega \in \mathcal{X},\ \|\omega\| \leq R \} \\ &= R \inf \left\{ \left\| \frac{1}{R} x - \varphi(A^*A)\tilde{\omega} \right\| \ : \ \tilde{\omega} \in \mathcal{X},\ \|\tilde{\omega}\| \leq 1 \right\} \\ &= \frac{1}{t} \hat{d}_\varphi(t;x) \end{aligned}$$

with $t := \frac{1}{R}$. We cite the following proposition (see [46, Theorem 5.9]).

Proposition 2.4.1 *Let $x \notin \mathcal{R}(\varphi(A^*A))$ but $x = \psi(A^*A)\omega$ for some $\omega \in \mathcal{X}$ with $\|\omega\| \leq K$. Moreover, the quotient $\left(\frac{\varphi}{\psi}\right)(t)$ is an index function for $0 < t \leq \|A\|^2$. Then we have the estimate*

$$\hat{d}_\varphi(t;x) \leq \varphi\left(\left(\frac{\varphi}{\psi} \right)^{-1} (K\,t) \right) \quad \textit{for all} \quad 0 < t \leq \frac{1}{K} \frac{\varphi(\|A\|^2)}{\psi(\|A\|^2)}.$$

It is clear, that $\left(\frac{\varphi}{\psi}\right)(t) \to 0$ as $t \to 0$, see [46, Lemma 5.8]. So, the additional condition $\left(\frac{\varphi}{\psi}\right)(t)$ being an index function is not very restrictive. We rewrite the result to our notation.

Corollary 2.4.1 *Under the condition of Proposition 2.4.1 we have the estimate*

$$d_\varphi(R;x) \le K\,\psi\left(\left(\frac{\psi}{\varphi}\right)^{-1}\left(\frac{R}{K}\right)\right) \quad \textit{for all} \quad R \ge K\frac{\psi(\|A\|^2)}{\varphi(\|A\|^2)}. \tag{2.9}$$

Proof. For $0 < t \le K^{-1}\varphi(\|A\|^2)\psi(\|A\|^2)^{-1}$ let $u \in \mathbb{R}$ be chosen such that $K\,t = \frac{\varphi(u)}{\psi(u)}$ or, equivalently, $u = \left(\frac{\varphi}{\psi}\right)^{-1}(K\,t)$. Moreover, for $R = \frac{1}{t}$ we derive

$$\frac{R}{K} = \frac{1}{K\,t} = \frac{\psi(u)}{\varphi(u)} \Leftrightarrow u = \left(\frac{\psi}{\varphi}\right)^{-1}\left(\frac{R}{K}\right).$$

Then we can write

$$\begin{aligned} d_\varphi(R;x) = \frac{1}{t}\hat{d}_\varphi(t;x) &\le R\,\varphi\left(\left(\frac{\varphi}{\psi}\right)^{-1}(K\,t)\right) \\ &= R\,\varphi(u) \\ &= K\frac{\psi(u)}{\varphi(u)}\varphi(u) = K\,\psi\left(\left(\frac{\psi}{\varphi}\right)^{-1}\left(\frac{R}{K}\right)\right). \end{aligned}$$

This proves the estimate. ■

With support of this corollary we are now able to compare the estimates (2.3) and (2.7) when we replace the function $\varphi(t)$ by $\psi(t)$ and R by K in the first estimate. Therefore we replace the distance function by its upper bound (2.9) and repeat the first balancing step of the proof of Theorem 2.4.1. We obtain

$$\varphi(\alpha) = \frac{K}{R}\psi\left(\left(\frac{\psi}{\varphi}\right)^{-1}\left(\frac{R}{K}\right)\right) = \frac{\varphi(u)}{\psi(u)}\psi(u) = \varphi(u)$$

by assuming R sufficiently large and u chosen is in the proof of Corollary 2.4.1. Hence $u = \alpha$ holds since $\varphi(t), t \ge 0$, is strictly increasing. Moreover, we observe for R sufficiently large, that

$$\begin{aligned} (C_\varphi + C_2)\,d\left(\Theta^{-1}(\alpha)\right) &= C_\varphi\varphi(\alpha)\,R + C_2 d_\varphi(R;x^\dagger) \\ &\le C_\varphi\varphi(\alpha)\,R + C_2 K\,\psi\left(\left(\frac{\psi}{\varphi}\right)^{-1}\left(\frac{R}{K}\right)\right) \\ &= C_\varphi\varphi(\alpha)K\frac{\psi(\alpha)}{\varphi(\alpha)} + C_2 K\,\psi(\alpha) \\ &= (C_\varphi + C_2)\,K\,\psi(\alpha). \end{aligned}$$

Hence, estimate (2.7) is at least of same order than estimate (2.3). We give the example of power-type and logarithmic source conditions.

Example 2.4.1 (Power-type source conditions) *We consider the case of power-type source conditions. Assume $x^\dagger \in \mathcal{R}((A^*A)^\nu)$ but $x^\dagger \notin \mathcal{R}((A^*A)^\mu)$ for two exponents $\mu, \nu > 0$. Then, necessarily, $\mu > \nu$ holds. As estimate for the distance function with $\psi(t) = t^\nu$ and $\varphi(t) = t^\mu$ we derive*

$$d_\varphi(R; x^\dagger) \le C\, R^{\frac{\nu}{\nu-\mu}}, \quad R \ge R_0$$

for some constant $C > 0$ and $R_0 > 0$ sufficiently large. In example, for $\mu = \frac{1}{2}$, i.e. $x^\dagger \notin \mathcal{R}(A^*)$ and $\nu < \frac{1}{2}$ we derive $d_\varphi(R; x^\dagger) \le C\, R^{\frac{2\nu}{2\nu-1}}$ which is the result in [44, Theorem 1] for compact operators A. The case $\mu = 1$ was considered in [17, Theorem 3.2] again in the compact case. As opposite to the general result here, no lower bound on the parameter R is needed in this specific situation. So the additional bound seems to have more technical reasons than it is a really restrictive one. This becomes more evident by the observation that the lower bound vanishes when we replace $\|A\|^2$ in the Definitions 2.4.1 and 2.4.3 by an arbitrary large bound (and noticing that $\frac{\psi(t)}{\varphi(t)} \to \infty$ as $t \to \infty$).

Example 2.4.2 (Logarithmic source conditions) *Besides power-type source conditions the examination of logarithmic source conditions has been well-established in the recent years especially for severely ill-posed problems, see e.g. [49] and [50]. For given $x^\dagger \in \mathcal{X}$, the source condition $x^\dagger \in \mathcal{R}(\psi(A^*A))$ with $\psi(t) := \frac{1}{[\log(t^{-1})]^\mu}$, $t < 1$, for some $\mu > 0$ was assumed. Additionally we suppose that $x^\dagger \notin \mathcal{R}((A^*A)^{\frac{1}{2}})$. It seems to be difficult to transfer the concept of logarithmic source conditions directly into the theory of distance functions. We therefore give two approaches:*

- *We can apply Corollary 2.4.1 with $\psi(t) := \frac{1}{[\log(t^{-1})]^\mu}$, $\mu > 0$, and $\varphi(t) = t^{\frac{1}{2}}$, $t \ge 0$. For simplicity we assume $K = 1$. Then the corresponding distance function $d(R) = d_\varphi(R; x^\dagger)$ satisfies*

$$\sqrt{\alpha} R(\alpha) = d(R(\alpha)) = \frac{1}{[\log(\alpha^{-1})]^\mu}$$

for some proper chosen function $R = R(\alpha)$. We derive

$$R(\alpha) = \frac{1}{\sqrt{\alpha}[\log(\alpha^{-1})]^\mu} =: \tilde{\Theta}(\alpha) \Leftrightarrow \alpha = \tilde{\Theta}^{-1}(R).$$

Hence the corresponding distance function is given as

$$d(R) = \frac{1}{\left[\log\left(\frac{1}{\tilde{\Theta}^{-1}(R)}\right)\right]^\mu} = R\sqrt{\tilde{\Theta}^{-1}(R)}.$$

It is interesting to observe that obtain two representations of the distance function. Both versions coincide which might not be obviously on the first glance. On the other hand both variants contain the function $\tilde{\Theta}^{-1}(R)$, $R > 0$, which cannot specified explicitly. So we have no explicit representation of the distance function which describes a logarithmic source condition (for the accordant reference source condition under consideration).

- *We follow an alternative approach given e.g. in [40, Example 4.7]. Here a logarithmic decay of the source condition is assumed explicitly, i.e.*

$$d(R) = d_\varphi(R; x^\dagger) = \frac{C}{[\log R]^\mu}, \qquad R > 1,$$

holds for some constant $C > 0$ and given $x^\dagger \in \mathcal{X}$. It can be shown that this condition is in fact somewhat stronger than the assumption of a logarithmic source condition. Choosing $R(\alpha) = \alpha^{-\nu}$ for some $0 < \nu < \frac{1}{2}$, the term $\sqrt{\alpha}R(\alpha) = \alpha^{\frac{1}{2}-\nu}$ decays faster to zero as $\alpha \to 0$ as the term

$$d(R(\alpha)) = \frac{C}{[\log R(\alpha)]^\mu} = \frac{C}{\nu^\mu[\log(\alpha^{-1})]^\mu}$$

which is of same order than in the first case where the logarithmic source condition for $x^\dagger \in \mathcal{X}$ was assumed. Since the choice of $R = R(\alpha)$ in this case was not optimal keeping the terms $\sqrt{\alpha}R(\alpha)$ and $d(R(\alpha))$ of same order, the decay rate of the distance function here is faster. Such faster decay of the distance function is in fact a stronger (smoothness) condition on the element $x^\dagger \in \mathcal{X}$.

We now deal with the opposite question: for $x^\dagger \in \mathcal{R}(A^\dagger)$ and some index function $\varphi(t)$, $t \geq 0$, let the distance function $d(R) = d_\varphi(R; x^\dagger)$ be given. So, can we draw conclusions about the validity of a general source condition $x^\dagger \in \mathcal{R}(\psi(A^*A))$ for some index function $\psi(t)$, $t \geq 0$? This question is still not answered in general. Based on the converse result [76, Corollary 2.6] we can find the following statement for power-type distance functions, see also [17, Corollary 3.3].

Proposition 2.4.2 *For $0 < \mu \leq 1$ we assume $\varphi(t) = t^\mu$, $t \geq 0$ and $x^\dagger \in \mathcal{R}(A^\dagger)$ with $x^\dagger \notin \mathcal{R}((A^*A)^\mu)$ has a distance function satisfying $d_\varphi(R; x^\dagger) \leq C\, R^{\frac{\nu}{\nu-\mu}}$ for some constant $C > 0$ and $0 < \nu < \mu$. Then $x^\dagger \in \mathcal{R}\left((A^*A)^{\tilde{\nu}}\right)$ holds for all $0 < \tilde{\nu} < \nu$.*

Comparing the results of Example 2.4.1 and Proposition 2.4.2 we observe a gap in the limit situation: using the notation Proposition 2.4.2 we can suppose a distance function $d_\varphi(R) \leq C\, R^{\frac{\nu}{\nu-\mu}}$ if we additionally assume that $x^\dagger \in \mathcal{R}((A^*A)^\nu)$ with $0 < \nu < \mu$. On the other hand, from the distance function $d_\varphi(R) \leq C\, R^{\frac{\nu}{\nu-\mu}}$ we only can collude that $x^\dagger \in \mathcal{R}\left((A^*A)^{\tilde{\nu}}\right)$ holds for all $0 < \tilde{\nu} < \nu$. In particular, we cannot conclude $x^\dagger \in \mathcal{R}((A^*A)^\nu)$, which might be the first idea. We present an example that shows that this gap really exists.

Example 2.4.3 *We return to our standard example 1.2.1 with $p = q = 2$. We set $\varphi(t) := t^{\frac{1}{2}}$, $t \geq 0$. Moreover, let $x^\dagger \equiv 1$. Then from [36, Example 4.5] we conclude $d(R; x^\dagger) \leq C\, R^{-1} = C\, R^{\frac{\nu}{\nu-1/2}}$, $R \geq 1$, with $\nu = \frac{1}{4}$ and some constant $C > 0$. On the other hand, the calculation in [44, Example 4] evidences that $x^\dagger \in \mathcal{R}\left((A^*A)^{\tilde{\nu}}\right)$ for all $0 < \tilde{\nu} < \frac{1}{4}$. This observation proves two points: first, the estimate of the distance functions is of correct order, i.e. the exponent cannot be decreased. The second observation is more remarkable. Based on distance functions we can prove for $x^\dagger$ the optimal convergence rate $\|x_\alpha^\delta - x^\dagger\| = \mathcal{O}\left(\delta^{\frac{1}{3}}\right)$ as $\delta \to 0$ by choosing $\alpha := \delta^{\frac{4}{3}}$. However, this rate cannot be proved with the aid of general source conditions.*

This example evidences that this limit situation does not have only technical reasons based on estimates in the proofs. Moreover, such limit situation was also observed in [47, Corollary 1] for general source conditions: each $x^\dagger \in \mathcal{R}(A^\dagger)$ has either unlimited smoothness, i.e. $x^\dagger \in \mathcal{R}(\psi(A^*A))$ for all index functions $\psi(t)$, $t \geq 0$, or there exists an index function $\bar{\psi}(t)$, $t \geq 0$ with $x^\dagger \notin \mathcal{R}(\bar{\psi}(A^*A))$ but $x^\dagger \in \mathcal{R}(\psi(A^*A))$ for all index functions $\psi(t)$, $t \geq 0$ satisfying $\frac{\bar{\psi}(t)}{\psi(t)} \to 0$ as $t \to 0$. If $\bar{\psi}(t)$ is a qualification of the regularization $\{g_\alpha\}$ so we cannot prove the estimate (2.3) with φ replaced by $\bar{\psi}$ and an accordant convergence rate (but any weaker). On the other hand, the considerations above has shown that if $\bar{\psi}(t)$ is of power-type we can prove such convergence rate based on the concept of distance functions. So we can formulate here only the following open question:

- Assume $x^\dagger \notin \mathcal{R}(\varphi(A^*A))$ has distance function $d_\varphi(R; x^\dagger)$, $R \geq 0$. Moreover, there exists an index function $\psi(t)$, $t \geq 0$, such that the estimate (2.9) of Corollary 2.4.1 holds for some constant $K > 0$. So, can we then conclude that $x^\dagger \in \mathcal{R}\left(\tilde{\psi}(A^*A)\right)$ for all index functions $\tilde{\psi}(t)$, $t \geq 0$, with $\frac{\psi(t)}{\tilde{\psi}(t)} \to 0$ as $t \to 0$?

The concept of distance functions seems to be more advantageous in this limit situation at least in the case of power-type distance functions based on a power-type reference source condition.

Finally we present an alternative proof of Lemma 1.1.1 based on approximative source condition.

Second proof of Lemma 1.1.1. Since $x^\dagger \in \mathcal{R}((A^*A)^\mu)$ we have for the distance function $d(R) = d(R; x^\dagger) \leq C\, R^{\frac{2\mu}{2\mu-1}}$ for some constant $C > 0$. Hence, for each $R > 0$ we can find a representation $x^\dagger = A^*\omega(R) + v(R)$ with $\|\omega(R)\| \leq R$ and $\|v(R)\| \leq d(R)$. Hence we can continue

$$\begin{aligned} \|A\, x_\alpha^\delta - y^\delta\| + \frac{\alpha}{2}\|x_\alpha^\delta - x^\dagger\|^2 &\leq \delta + \alpha \left\langle A^*\omega(R) + v(R), x_\alpha^\delta - x^\dagger \right\rangle_{\mathcal{X}} \\ &\leq \delta + \alpha\|\omega(R)\| \left(\|A\, x_\alpha^\delta - y^\delta\| + \delta\right) + \alpha\|v(R)\|\|x_\alpha^\delta - x^\dagger\| \\ &\leq \delta + \alpha\, R \left(\|A\, x_\alpha^\delta - y^\delta\| + \delta\right) + \alpha\, d(R)\|x_\alpha^\delta - x^\dagger\|. \end{aligned}$$

This inequality holds for all $R > 0$. We now set $R := \frac{1}{\alpha}$. Then we arrive at

$$\|x_\alpha^\delta - x^\dagger\|^2 \leq \frac{4\,\delta}{\alpha} + C\,\alpha^{\frac{2\mu}{1-2\mu}}\|x_\alpha^\delta - x^\dagger\|,$$

which again implies

$$\|x_\alpha^\delta - x^\dagger\| \leq 2\frac{\sqrt{\delta}}{\sqrt{\alpha}} + C\,\alpha^{\frac{2\mu}{1-2\mu}}.$$

The final balancing step is the same as in the first proof. ■

This example shows the basic idea of the application of the technique of distance functions in the next chapter. With respect to the general Banach space setting with general stabilizing functional P we only have to replace the used equality

$$\frac{1}{2}\|x_\alpha^\delta\|^2 - \frac{1}{2}\|x^\dagger\|^2 = \frac{1}{2}\|x_\alpha^\delta - x^\dagger\|^2 + \langle x^\dagger, x_\alpha^\delta - x^\dagger\rangle_{\mathcal{X}}$$

by its generalization

$$P(x_\alpha^\delta) - P(x^\dagger) = D_{\xi^\dagger}(x_\alpha^\delta, x^\dagger) + \langle \xi^\dagger, x_\alpha^\delta - x^\dagger \rangle$$

where $\xi^\dagger \in \partial P(x^\dagger)$. In fact, with $\mathcal{X}$ being a Hilbert space and $P(x) = \frac{1}{2}\|x\|^2$ the second equation automatically reduces to the first one. This observation also motivates the use of Bregman distances in the subsequent error analysis.

Chapter 3

Tikhonov regularization in Banach spaces

3.1 Preliminaries

3.1.1 Basic assumptions

Recalling the notations already employed in the introduction we deal with the following assumptions, see e.g. [45]:

(A1) The spaces $\mathcal{X}$ and $\mathcal{Y}$ are assumed to be real Banach spaces and $F : \mathcal{D}(F) \subseteq \mathcal{X} \longrightarrow \mathcal{Y}$ denotes a nonlinear operator with domain $\mathcal{D}(F)$. Moreover, the operator F is continuous with respect to the weak topologies and $\mathcal{D}(F)$ is weakly sequentially closed.

(A2) The stabilizing functional $P : \mathcal{X} \longrightarrow [0, \infty]$ is non-negative, convex and weakly lower semi-continuous with domain $\mathcal{D}(P) := \{x \in \mathcal{X} : P(x) < \infty\}$ and the set $\mathcal{D} := \mathcal{D}(F) \cap \mathcal{D}(P)$ is nonempty.

(A3) We have $1 \leq p < \infty$, $\delta \in [0, \delta_{max}]$, $\alpha = \alpha(\delta) \in [\alpha_{min}(\delta), \alpha_{max}]$ for given lower bound $\alpha_{min}(\delta) \geq 0$ with $\alpha_{min}(\delta) \to 0$ for $\delta \to 0$ and $\delta^p/\alpha_{min}(\delta)$ remains bounded on $[0, \delta_{max}]$.

(A4) For every $0 < \alpha \leq \alpha_{max}$ and $c \geq 0$ the sets

$$\mathcal{S}_{\alpha,y}(c) := \left\{ x \in \mathcal{D} \, : \, \frac{1}{p}\|F(x) - y\|^p + \alpha\, P(x) \leq c \right\} \tag{3.1}$$

are weakly sequentially pre-compact, i.e. every sequence $\{x_k\} \subset \mathcal{S}_{\alpha,y^\delta}(c)$ has a subsequence which is weakly convergent in $\mathcal{X}$.

(A5) For exact data $y \in \mathcal{Y}$ there exists a P-minimizing solution $x^\dagger \in \mathcal{D}$ of equation $F(x) = y$, i.e

$$F(x^\dagger) = y \quad \text{and} \quad P(x^\dagger) = \min\{P(x) \, : \, F(x) = y\}$$

holds. Moreover, we suppose the existence of $\xi^\dagger \in \partial P(x^\dagger)$ with $\xi^\dagger \neq 0$.

As already mentioned in the introduction we deal with stable approaches for solving the nonlinear ill-posed operator equation

$$F(x) = y, \qquad x \in \mathcal{D}(F), \tag{3.2}$$

approximately. We remember that for linear equations with $\mathcal{D}(F) = \mathcal{X}$ we will write $F = A$ and the nonlinear equation (3.2) reformulated as linear equation

$$A\,x = y, \qquad x \in \mathcal{X}. \tag{3.3}$$

Furthermore, we assume that we do not know the data $y \in \mathcal{Y}$ exactly. Only a noisy observation $y^\delta \in \mathcal{Y}$ is given. Here, $\delta \geq 0$ describes the information about the (absolute) size of the noise level, i.e. the estimate $\|y - y^\delta\| \leq \delta$ is supposed to be known.

The ill-posedness forces the application of regularization methods. Here in this chapter the focus is on Tikhonov regularization approaches: for given $\alpha > 0$ and data $y^\delta \in \mathcal{Y}$ we define the Tikhonov functional

$$T_{\alpha,y^\delta}(x) := \frac{1}{p}\|F(x) - y^\delta\|^p + \alpha\, P(x), \qquad x \in \mathcal{D}.$$

Instead trying to solve equation (3.2) exactly we now deal with the minimization problem

$$T_{\alpha,y^\delta}(x) \to \min \quad \text{subject to } x \in \mathcal{D}. \tag{3.4}$$

A solution of (3.4) we denote as usual with x_α^δ, which we call a (Tikhonov-)regularized approximate solution of equation (3.2).

We shortly give two remarks concerning conditions (A1) and (A4). In particular (A1) implies the following: for any sequence $\{x_n\} \subset \mathcal{D}(F)$ with $x_n \rightharpoonup x \in \mathcal{X}$ and $F(x_n) \rightharpoonup y \in \mathcal{Y}$ we can conclude $x \in \mathcal{D}(F)$ and $F(x) = y$. It is also well-known that a linear bounded operator $A : \mathcal{X} \longrightarrow \mathcal{Y}$ is also continuous with respect to the weak topologies on $\mathcal{X}$ and $\mathcal{Y}$, see [87, Lemma 8.49]. So, (A1) is automatically satisfied for linear equations (with bounded operator). For the weak closedness of $\mathcal{D}(F)$ it is sufficient that $\mathcal{D}(F)$ is convex and closed.

Furthermore, assume $x \in \mathcal{S}_{\alpha,y^\delta}(c) := \left\{x \in \mathcal{D} \,:\, T_{\alpha,y^\delta}(x) \leq c\right\}$ for some $c > 0$. Then

$$\begin{aligned}
\frac{1}{p}\|F(x) - y\|^p + \alpha\, P(x) &\leq 2^{p-1}\left(\frac{\delta^p}{p} + \frac{1}{p}\|F(x) - y^\delta\|^p\right) + \alpha\, P(x) \\
&\leq 2^{p-1}\left(\frac{\delta_{max}^p}{p} + \frac{1}{p}\|F(x) - y^\delta\|^p + \alpha\, P(x)\right) \\
&\leq 2^{p-1}\left(\frac{\delta_{max}^p}{p} + c\right).
\end{aligned}$$

Hence $x \in \mathcal{S}_{\alpha,y}(\tilde{c})$ with $\tilde{c} := 2^{p-1}(\delta_{max}^p p^{-1} + c)$ which shows that also the sets $\mathcal{S}_{\alpha,y^\delta}(c)$ are weakly sequentially pre-compact for arbitrary $y^\delta \in \mathcal{B}_{\delta_{max}}(y)$.

Later on, for proving convergence rates results we will state further assumptions such as non-linearity restrictions to the operator F and smoothness conditions to the stabilizing functional P.

3.1.2 Stability and convergence results

Since our focus is here on the convergence rates analysis we shortly recall the main results concerning stability and convergence of the regularization approach (3.4). We cite the following result, see [45, Theorem 3.1, 3.2 and 3.5].

Proposition 3.1.1 *Assume (A1)-(A5). Then the following holds true:*

(i) There exists a minimizer $x_\alpha^\delta \in \mathcal{D}$ *of (3.4).*

(ii) If $\{y_k\}$ *with* $y_k \to y^\delta$*, then every sequence* $\{x_k\}$ *of solutions of (3.4) with* y^δ *replaced by* y_k *has a weakly convergent subsequence. Each limit of such subsequence* $\{x_{k_m}\}$ *is a minimizer of (3.4). Moreover* $P(x_{k_m})$ *converges to* $P(x_\alpha^\delta)$*.*

(iii) Assume $\delta_k \to 0$, $y_k = y^{\delta_k}$ *satisfies* $\|y - y_k\| \le \delta_k$ *and* $\alpha_k = \alpha(\delta_k)$ *satisfies*

$$\alpha(\delta) \to 0 \quad \text{and} \quad \frac{\delta^p}{\alpha(\delta)} \to 0 \quad \text{as} \quad \delta \to 0.$$

Then, every sequence $\{x_{\alpha_k}^{\delta_k}\}$ *of solutions of (3.4) with* y^δ *replaced by* y_k *and* α *replaced by* α_k *has a weakly convergent subsequence. Each limit of such subsequence* $\{x_{\alpha_{k_m}}^{\delta_{k_m}}\}$ *is a P-minimizing solution of (3.2). If in addition the P-minimizing solution* $x^\dagger$ *is unique, then* $x_{\alpha_k}^{\delta_k} \rightharpoonup x^\dagger$*.*

We point out, that the conditions (A1)-(A5) only ensure weak convergence in Proposition 3.1.1. In order to prove strong convergence we need an additional assumption. Such condition can be given for example via

$$\{x_n\} \subset \mathcal{D}(P),\ x_n \rightharpoonup x \in \mathcal{D}(P),\ P(x_n) \to P(x) \implies x_n \to x, \tag{3.5}$$

see e.g. [89]. Then the following holds immediately.

Corollary 3.1.1 *Assume (A1)-(A5). If additionally (3.5) holds then the convergence in Proposition 3.1.1(ii) and (iii) is strong.*

An alternative condition to the functional P ensuring strong convergence $x_\alpha^\delta \to x^\dagger$ (at least for subsequences) can be found e.g. in [87]: instead of (3.5) total convexity of P is supposed, see e.g. [87, Definition 3.29 + Proposition 3.32].

We also frequently make use of the following result, see also [45, Remark 3.6].

Lemma 3.1.1 *Assume (A1)-(A5),* $y^\delta \in \mathcal{B}_{\delta_{max}}(y)$*, and* $x_\alpha^\delta \in \mathcal{D}$ *is a solution of (3.4). Then there exist two constants* $\varrho_y, \varrho_x > 0$ *(not depending on* α *and* δ*) such that*

$$\|F(x_\alpha^\delta) - y^\delta\| \le \varrho_y \quad \text{and} \quad \|x_\alpha^\delta - x^\dagger\| \le \varrho_x.$$

PROOF. Since $T_{\alpha,y^\delta}(x_\alpha^\delta) \le T_{\alpha,y^\delta}(x^\dagger)$ we have

$$\|F(x_\alpha^\delta) - y^\delta\| \le \left(\delta^p + p\,\alpha P(x^\dagger)\right)^{\frac{1}{p}} \le \left(\delta_{max}^p + p\,\alpha_{max} P(x^\dagger)\right)^{\frac{1}{p}} =: \varrho_y$$

and $P(x_\alpha^\delta) \le \frac{1}{p}\delta^p/\alpha + P(x^\dagger)$. We conclude

$$\begin{aligned}
T_{\alpha_{max},y^\delta}(x_\alpha^\delta) &\le \frac{1}{p}\|F(x_\alpha^\delta) - y\|^p + \alpha_{max} P(x_\alpha^\delta) \\
&\le 2^{p-1}\left(\frac{1}{p}\|F(x_\alpha^\delta) - y^\delta\|^p + \alpha_{max} P(x_\alpha^\delta) + \frac{\delta^p}{p}\right) \\
&= 2^{p-1}\left(\frac{1}{p}\|F(x_\alpha^\delta) - y^\delta\|^p + \alpha P(x_\alpha^\delta) + (\alpha_{max} - \alpha) P(x_\alpha^\delta) + \frac{\delta^p}{p}\right) \\
&\le 2^{p-1}\left(\frac{1}{p}\|F(x^\dagger) - y^\delta\|^p + \alpha P(x^\dagger) + (\alpha_{max} - \alpha) P(x_\alpha^\delta) + \frac{\delta^p}{p}\right) \\
&\le 2^{p-1}\alpha_{max}\left(P(x^\dagger) + \frac{2\delta^p}{p\,\alpha}\right).
\end{aligned}$$

Here we used that

$$(\alpha_{max} - \alpha) P(x_\alpha^\delta) \le (\alpha_{max} - \alpha)\left(\frac{\delta^p}{p\,\alpha} + P(x^\dagger)\right) \le \alpha_{max} P(x^\dagger) + \alpha_{max}\frac{\delta^p}{p\,\alpha} - \alpha P(x^\dagger) - \frac{\delta^p}{p}.$$

By assumption (A4) the term in the bracket remains bounded, i.e. there exists

$$C := \sup_{\delta \in [0,\delta_{max}]} \delta^p \alpha^{-1} \quad \text{and hence} \quad \varrho_{max} := 2^{p-1}\alpha_{max}(P(x^\dagger) + 2Cp^{-1}).$$

Hence $x_\alpha^\delta, x^\dagger \in \mathcal{S}_{\alpha_{max},y}(\varrho_{max})$. Since the set is weakly sequentially pre-compact it is bounded. In particular, there exists a uniform constant $\varrho_x > 0$ such that $\|x_\alpha^\delta - x^\dagger\| \le \varrho_x$ holds. ∎

We make use of these constants ϱ_{max}, ϱ_x and ϱ_y later on in our considerations.

3.1.3 Some prototypical applications

Before we are going on with our discussion we briefly recall some exampels of penalty functionals P which have been well-established in a various number of applications in the recent years.

a) Quadratic penalty terms

We suppose $\mathcal{X}$ to be a Hilbert space. Moreover, let $L : \mathcal{D}(L) \subseteq \mathcal{X} \longrightarrow \mathcal{Z}$ a closed, densely defined linear operator with $\mathcal{R}(L) = \mathcal{Z}$. Here, $\mathcal{Z}$ describes another Hilbert space. We now introduce the quadratic penalty functional $P : \mathcal{X} \longrightarrow [0,\infty]$ as

$$P(x) := \begin{cases} \frac{1}{2}\|L\,x\|^2, & x \in \mathcal{D}(L), \\ \infty, & \text{else.} \end{cases} \tag{3.6}$$

This situation was originally examined in [66] for linear operator equations in Hilbert spaces. Later on, the basic ideas of this regularization approach could be generalized to arbitrary linear regularization methods, see e.g. [21, Chapter 8]. In particular, in [26] iterative regularization methods were considered. Moreover, in [60] we find a convergence analysis (including a convergence rates result) for Tikhonov regularization of nonlinear equations (again in Hilbert spaces).

Example 3.1.1 *One of the main application are differential operators. Let* $\mathcal{X} = \mathcal{Z} = L^2(0,1)$ *and* $\mathcal{D}(L) := H^l(0,1)$ *for some* $l \in \mathbb{N}$. *Then we set* $L\,x := x^{(l)}$. *In particular the choice* $l = 2$ *has been well-established in many practical applications. This example also shows the need of dealing with closed operators (in particular with discontinuous operators) owning a non-trivial null space.*

Therefore, this method is often called *regularization with differential operators* or *regularization with semi-norms.* We present also another application.

Example 3.1.2 *Another approach is the so-called* regularization in Hilbert scales *which was introduced in [75]. Here,* $\mathcal{Z} = \mathcal{X}$ *is chosen and* $B : \mathcal{X} \longrightarrow \mathcal{X}$ *is supposed to be a densely defined unbounded selfadjoint and strictly positive operator. For given* $s \in \mathbb{R}$ *we set* $L := B^s$. *We refer to [21, Sections 8.4. and 8.5.] for more detailed information about this topic.*

The following important result can be found in [89, Lemma 1].

Lemma 3.1.2 *Under the condition stated above the functional* P *defined in (3.6) is convex and weakly lower semi-continuous.*

In particular, the functional P satisfies (A2) as long as $\mathcal{D}(F)$ is nonempty. In order to ensure (A4) we need conditions which connect properties of the operator F and the functional P. In [89] we find the following:

- For $\{x_k\} \subset \mathcal{D}(F)$ we set $x_k := x_k^0 + x_k^1$ with $x_k^0 \in \mathcal{N}(L)$ and $x_k^1 \in \mathcal{N}(L)^\perp$. Then, if the sequence $\{F(x_k)\}$ is bounded in $\mathcal{Y}$ we suppose that $\{x_k^0\}$ is bounded.

This condition clearly implies (A4). A slightly different condition can be found in [60]. We also point out, that these conditions only ensures weak convergence in $\mathcal{X}$. In order to obtain strong convergence we need an additional assumption. We cite the following lemma, see [60, Lemma 1].

Lemma 3.1.3 *Let* $L : \mathcal{D}(L) \subseteq \mathcal{X} \longrightarrow \mathcal{Z}$ *a closed linear operator between the Hilbert spaces* $\mathcal{X}$ *and* $\mathcal{Z}$. *Then the following conditions are equivalent:*

(i) If $x_n \rightharpoonup x$ *and* $\|L\,x_n\| \to \|L\,x\|$ *for a sequence* $\{x_n\} \subset \mathcal{D}(L)$, *then* $x_n \to x$.

(ii) $\mathcal{N}(L)$ *is finite dimensional and* $\mathcal{R}(L)$ *is closed.*

Hence, *(ii)* gives a necessary and sufficient condition for the validity of assumption (3.5) yielding stability and convergence of the regularization approach (3.4) with respect to the norm topology in $\mathcal{X}$.

Finally we discuss differentiability. For all $x, h \in \mathcal{D}(L)$ we easily derive the directional derivative

$$P'(x)h := \lim_{\varepsilon \to 0} \frac{P(x + \varepsilon\, h) - P(x)}{\varepsilon} = \langle L\, x, L\, h\rangle.$$

Hence, P is Fréchet-differentiable for all $x \in \mathcal{D}(L)$ satisfying $L\, x \in \mathcal{D}(L^*)$ and

$$P'(x) = L^* L\, x \in \mathcal{X}.$$

Here $L^* : \mathcal{D}(L^*) \subseteq \mathcal{Z} \longrightarrow \mathcal{X}$ denotes the (Hilbert-)adjoint operator of L and we identify the space $\mathcal{X}$ with its dual $\mathcal{X}^*$. Hence, condition $P'(x^\dagger) \neq 0$ of (A5) implies $L\, x^\dagger \notin \mathcal{N}(L)$.

Remark 3.1.1 *In the theory of nonlinear equations in Hilbert spaces often the regularization term $P(x) := \frac{1}{2}\|x - x^*\|^2$ for some given a-priori guess $x^* \in \mathcal{X}$ is considered, see e.g. [22].*

b) Bounded variation penalization

Here, we follow the concept of [1]. Therefore, let $\Omega \subset \mathbb{R}^d$, $d \in \mathbb{N}$, be a bounded convex domain with Lipschitz continuous boundary. We define

$$\mathcal{V} := \left\{ v \in C_0^1(\Omega; \mathbb{R}^d) \;:\; |v(t)| \leq 1\; \forall\, t \in \Omega \right\}$$

as set of test functions. Here, $|v| := \sqrt{\sum_{j=1}^{d} v_j^2}$, $v = (v_1, \ldots, v_d)^T \in \mathbb{R}^d$ denotes the Euclidean norm on $\mathbb{R}^d$. Then, for $\beta \geq 0$ we define

$$J_\beta(x) := \sup_{v \in \mathcal{V}} \int_\Omega \left(-x \operatorname{div} v + \sqrt{\beta(1 - |v(t)|^2)} \right) dt.$$

Moreover, the space of functions of bounded variation is now defined as

$$\mathrm{BV}(\Omega) := \left\{ x \in L^1(\Omega) \;:\; J_0(x) < \infty \right\}.$$

Together with the norm $\|x\|_{\mathrm{BV}} := \|x\|_{L^1} + J_0(x)$ the space $\mathrm{BV}(\Omega)$ becomes a Banach space. We can also cite the following embedding theorem, see [2, Corollary 3.49].

Proposition 3.1.2 *The space $BV(\Omega)$ is compactly embedded in $L^p(\Omega)$ for $1 \leq p < d/(d-1)$. For $d \geq 2$, $BV(\Omega)$ is continuously embedded in $L^{\frac{d}{d-1}}(\Omega)$.*

We summarize the most important properties, see [1].

Proposition 3.1.3 *The following holds true:*

(i) For any $\beta > 0$ and $x \in L^1(\Omega)$ we have $J_0(x) < \infty$ if and only if $J_\beta(x) < \infty$ and $\lim_{\beta \to 0} J_\beta(x) = J_0(x)$ if $x \in BV(\Omega)$.

(ii) If $x \in W^{1,1}(\Omega)$ then

$$J_\beta(x) = \int_\Omega \sqrt{|\nabla x|^2 + \beta}\, dt.$$

(iii) For $1 \le p < \infty$ we define the functional $P : L^p(\Omega) \longrightarrow [0, \infty]$ as

$$P(x) := \begin{cases} J_\beta(x), & x \in BV(\Omega), \\ \infty, & \text{else.} \end{cases}$$

Then the functional P is convex and weakly lower semi-continuous.

In particular, assumption (A2) is fulfilled for all L^p-spaces with $1 \le p < \infty$. The motivation of introducing β as additional parameter is quite simple: assume x to be sufficiently smooth, i.e. $x \in C^1(\Omega)$. Then, the functional J_0 fails to be differentiable in x whenever has local extremas in Ω, i.e. $\nabla x(t) = 0$ for some $t \in \Omega$. On the other hand, for $\beta > 0$, J_β is Gâteaux-differentiable for all $x \in \mathrm{BV}(\Omega)$. So, solving the underlying minimization problem, it easier to deal with J_β for positive β. We remark, that for $x, h \in C^1(\Omega)$, the directional derivative $P'(x)\, h$ is given as

$$P'(x)\, h := \int_\Omega \frac{\nabla x \cdot \nabla h}{\sqrt{|\nabla x|^2 + \beta}}\, dt.$$

It seems to be rather difficult for giving an explicit expression for $P'(x) \in \mathcal{X}^*$. Since we later on need $P'(x^\dagger)$ in source conditions for deriving our convergence rates results it will be hard to check such source conditions in practical applications.

Verifying assumption (A4) we observe the following: let $\{x_n\} \subset L^p(\Omega)$ be any sequence of constant functions with $\|x_n\| \to \infty$ as $n \to \infty$. Then we have obviously $P(x_n) \equiv \sqrt{\beta}\mathrm{meas}(\Omega)$ for all $n \in \mathbb{N}$. Hence we necessarily need the condition that $\|F(x_n)\| \to \infty$, see also [1, Lemma 4.1] in the case of linear operator equations. Then, by the embedding we can conclude that the level sets $\mathcal{S}_{\alpha,y}(c)$ are sequentially pre-compact for $1 \le p < d/(d-1)$ and weakly sequentially pre-compact for $p = d/(d-1)$ for $d \ge 2$.

We examine (3.5). Therefore, we assume $x_n \rightharpoonup x$ in $\mathcal{X} = L^p(\Omega)$, $1 \le p < d/(d-1)$, and $J_0(x_n) \to J_0(x)$. Then $x_n \rightharpoonup x$ also in $L^1(\Omega)$. Hence, the sequence $\{x_n\}$ is bounded in $\mathrm{BV}(\Omega)$. Hence there exists a convergent subsequence $x_{n_k} \to x$ in $L^p(\Omega)$ for $k \to \infty$. By the uniqueness of the weak limit of $\{x_n\}$ we can conclude the convergence of the whole sequence $x_n \to x$ in $L^p(\Omega)$ for $n \to \infty$.

Remark 3.1.2 *One might also think on choosing $\mathcal{X} = BV(\Omega)$ directly. Setting $P(x) := \|x\|_{BV}$ the weak lower semi-continuity is automatically satisfied since it is a general property of norms. On the other hand, if we set $P(x) := J_0(x)$ we have to replace the weak topology in (A2) and (A4) by the weak* topology of $BV(\Omega)$, see [87, Lemma 9.69]. We also point to [87, Lemma 9.68] for a characterization of the weak* convergence on $BV(\Omega)$.*

c) Maximum entropy regularization

This approach is widely studied in literature, first for linear operator equations, see e.g. [19], later on also for the nonlinear case, see e.g. [23]. Therefore we give here only a brief summary of the well-known results.

Assume $\Omega \subset \mathbb{R}^d$ to be a bounded domain and $\mathcal{X} = L^1(\Omega)$. Let the a-priori guess $x^* \in L^\infty(\Omega)$ satisfying

$$0 < C_1 \le x^*(t) \le C_2 < \infty \quad \text{a.e. on } \Omega \tag{3.7}$$

for two positive constants C_1 and C_2. Then we define the entropy functional

$$E(x, x^*) := \int\limits_\Omega \left[x(t) \log\left(\frac{x(t)}{x^*(t)}\right) - x(t) + x^*(t) \right] dt, \quad x \in \mathcal{X}. \tag{3.8}$$

Let $L^1_+(\Omega) \subset L^1(\Omega)$ denotes the set of all functions which are nonnegative a.e. on Ω. Furthermore we introduce the set

$$\mathcal{E}(\Omega) := \left\{ x \in L^1_+(\Omega) \, : \, \int_\Omega x \log x \, dt < \infty \right\}.$$

Then we have the following result.

Lemma 3.1.4 *For any x^* satisfying (3.7) we have $x \in \mathcal{E}(\Omega)$ if and only if $x \in L^1_+(\Omega)$ and $E(x, x^*) < \infty$.*

We define the functional $P : L^1(\Omega) \longrightarrow [0, \infty]$ as

$$P(x) := \begin{cases} E(x, x^*), & x \in \mathcal{E}(\Omega), \\ \infty, & \text{else.} \end{cases}$$

We summarize some properties of the functional P, see e.g. [19].

Lemma 3.1.5 *For any x^* satisfying (3.7) it holds:*

(i) $P(x) \ge P(x^*) = 0$ for all $x \in L^1_+(\Omega)$.

(ii) P is convex onto $L^1_+(\Omega)$ and strictly convex onto $\mathcal{E}(\Omega)$.

(iii) P is weakly lower semi-continuous in $L^1(\Omega)$.

(iv) The level sets $\{x : P(x) \le M\}$ are weakly compact for any constant $M > 0$.

(v) $\{x_n\} \subset \mathcal{E}(\Omega)$, $x \in \mathcal{E}(\Omega)$ with $P(x_n) \to P(x)$ and $x_n \rightharpoonup x$ in $L^1(\Omega)$ implies $x_n \to x$ in $L^1(\Omega)$.

This lemma shows the applicability of the entropy regularization. In particular we do not need a coercivity condition to the operator F. Consequently the weak closeness (assumption (A1)) is the only condition to the operator F for applying the maximum entropy regularization successfully as regularization method.

We consider the derivative. A short calculation shows for $x \in \mathcal{E}(\Omega)$ and $h \in L^1(\Omega)$, that the directional derivative is given as

$$P'(x)\, h := \lim_{\varepsilon \to 0} = \int\limits_\Omega \left(\log\left(\frac{x(t)}{x^*(t)}\right) - (x^*(t) - 1) \right) h(t) \, dt.$$

If $x \in L^\infty(\Omega) \cap \mathcal{E}(\Omega)$ and $\operatorname*{ess\,inf}_{t\in\Omega} x(t) > 0$ then P is Gâteaux differentiable in x and

$$P'(x) = \log\left(\frac{x}{x^*}\right) - (x^* - 1) \in \left(L^1(\Omega)\right)^* = L^\infty(\Omega).$$

For $x^* \equiv 1$ we have $P'(x) = \log x$ which was the case considered in [12].

d) General norms

We return to our general Banach space setting. We assume $\mathcal{X}$ to be a reflexive Banach space. Generalizing the classical version of Tikhonov regularization we set

$$P : \mathcal{X} \longrightarrow [0, \infty), \quad P(x) := \frac{1}{q}\|x\|^q, \quad x \in \mathcal{X},$$

for some parameter $q \geq 1$. Then, condition (A2) is automatically satisfied by the convexity and weak lower semi-continuity of the norm. Moreover, by definition, the sets $\mathcal{S}_{\alpha,y}(c)$ are bounded for all $c > 0$. A property of reflexive spaces is that bounded sets are always weakly sequentially pre-compact, see e.g. [41, Satz 60.6]. Hence (A4) is fulfilled without further assumptions. The additional condition (3.5) providing strong convergence of the regularization approach does not hold in general reflexive spaces. However, this condition holds if e.g.

- $\mathcal{X}$ is an arbitrary Hilbert space or
- we have $\mathcal{X} = L^p(\Omega)$, $\Omega \subset \mathbb{R}^d$, $1 < p < \infty$, see [55, Theorem 4.1.VIII].

We shortly discuss the choice of the parameter q. As long as $\mathcal{X}$ is a Hilbert space the (classical) choice $q = 2$ is natural and preferable. The further analysis in this chapter will show that it is sometimes reasonable to affiliate the choice with the convexity of the space $\mathcal{X}$: if $\mathcal{X}$ is supposed to be q-convex then we have from the Xu/Roach inequalities, see Corollary 2.2.1(ii), for $\xi = J_q(x^\dagger)$ that

$$D_\xi(x, x^\dagger) \geq \frac{C_q}{q}\|x - x^\dagger\|^q \Leftrightarrow \|x - x^\dagger\| \leq \left(\frac{q}{C_q} D_\xi(x, x^\dagger)\right)^{\frac{1}{q}}$$

holds for some constant $C_q > 0$. Since we will formulate convergence results with respect to Bregman distances these results can be easily transferred to convergence rates with respect to the norm.

Example 3.1.3 *Assume* $\mathcal{X} = L^p(\Omega)$, $\Omega \subset \mathbb{R}^d$, $1 < p < \infty$. *Then* $\mathcal{X}$ *is* $\max\{2, p\}$-*convex. Hence, the convexity implies the choice* $q := \max\{2, p\}$. *On the other hand, from the computational point the choice* $q = p$ *seems to be natural since it gives the duality mapping a more simple structure, see Example 2.2.1. Clearly, both choices coincides for* $p \geq 2$. *However, in particular the example of sparsity reconstruction suggested to choose* p *smaller than* 2. *In this case both choices* $q = 2$ *or* $q = p$ *makes sense.*

3.1.4 The degree of nonlinearity

We need a further assumption concerning the nonlinearity of the operator F.

(A6) For every $x \in \mathcal{D}$ there exists a $\varepsilon_0 > 0$ such that

$$x^\dagger + \varepsilon(x - x^\dagger) \in \mathcal{D} \quad \text{for all} \quad 0 \leq \varepsilon \leq \varepsilon_0.$$

Moreover, there exists a bounded linear operator $F'(x^\dagger) : \mathcal{X} \longrightarrow \mathcal{Y}$ such that

$$F'(x^\dagger)(x - x^\dagger) = \lim_{\varepsilon \to 0} \frac{1}{\varepsilon} \left(F(x^\dagger + \varepsilon(x - x^\dagger)) - F(x^\dagger) \right)$$

for all $x \in \mathcal{D}$.

With $F'(x^\dagger)^\star : \mathcal{Y}^* \longrightarrow \mathcal{X}^*$ we denote the dual operator of $F'(x^\dagger)$, e.g.

$$\langle y^*, F'(x^\dagger)\, x \rangle = \langle F'(x^\dagger)^\star y^*, x \rangle$$

holds for all $x \in \mathcal{X}$ and $y^* \in \mathcal{Y}^*$. The operator $F'(x^\dagger)$ can be considered as generalized Gâteaux-derivative of F at $x^\dagger \in \mathcal{D}(F)$. However, here we do not suppose that $x^\dagger$ is an interior point of $\mathcal{D}(F)$. As consequence, the operator $F'(x^\dagger)$ might not be uniquely determined. On the other hand, the uniqueness of such operator is not necessary in our error analysis. We also give the following example which gives reason for such generalization.

Example 3.1.4 *In fact, in many practical applications the interior of the domain $\mathcal{D}(F)$ can be empty even in Hilbert spaces. Let e.g. $\mathcal{X} = L^q(0,1)$, $1 \leq q < \infty$, and the domain of all admissible parameters $\mathcal{D}(F) := \{x \in \mathcal{X} \,:\, -\infty < a \leq x \leq b < \infty \;\text{ a.e. on } (0,1)\}$ contains lower and upper bounds on the co-domain of the functions belonging to $\mathcal{D}(F)$. Then the interior of $\mathcal{D}(F)$ is empty. Such conditions are e.g. often necessary for parameter identification problems in differential equations.*

Now we can define the degree of nonlinearity. Originally, it was introduced in [48] for nonlinear operators mapping in Hilbert spaces. Here, we adopt this concept and transfer the corresponding definition to the underlying Banach space and Bregman distance setting. We state the definition as follows, see also [40].

Definition 3.1.1 *Let $0 \leq c_1, c_2 \leq 1$ and $0 < c_1 + c_2 \leq 1$. We define F to be nonlinear of degree (c_1, c_2) for the Bregman distance $D_{\xi^\dagger}(\cdot, x^\dagger)$ of P at $x^\dagger \in \mathcal{D}$ and $\xi^\dagger \in \partial P(x^\dagger)$ if there exists a constant $L \geq 0$ such that*

$$\|F(x) - F(x^\dagger) - F'(x^\dagger)(x - x^\dagger)\| \leq L \|F(x) - F(x^\dagger)\|^{c_1} D_{\xi^\dagger}(x, x^\dagger)^{c_2} \tag{3.9}$$

hold for all $x \in \mathcal{S}_{\alpha_{max}, y}(\varrho_{max})$.

We are interested in the following situations:

- Case $L = 0$, i.e. the operator is linear. Then of course, the parameter c_1 and c_2 can be chosen arbitrarily. In fact, as we will see in some situations, the case of linear operators can be considered as special case of nonlinear operators with fixed choice of c_1 and c_2 but with $L = 0$.

- CASE $c_1 = 1$, $c_2 = 0$. Then the condition is also known as η-condition, see e.g. [28]. It was introduced for proving convergence rates for nonlinear Landweber iteration. There, the restriction $L < \frac{1}{2}$ was made. However no restrictions on L are made here a priorly.

- CASE $c_1 = 0$, $c_2 = 1$. This is the condition which was already considered in [82]. It generalizes the theory of [22] to Banach spaces. In particular, for Hilbert spaces $\mathcal{X}$ and $P(x) := \frac{1}{2}\|x - x^*\|^2$ we have $\xi^\dagger = P'(x^\dagger) = x^\dagger - x^*$ and $D_{\xi^\dagger}(x, x^\dagger) = \frac{1}{2}\|x - x^\dagger\|^2$. So, condition (3.9) coincides with the condition essentially used in the proof of the convergence rates result in [22].

- CASE $0 < c_1 < 1$. This weaken the strong connection of F and $F'(x^\dagger)$ in the second case. Such 'mixed' conditions can be also found for iterative regularization methods of Newton-type for nonlinear operator equations in Hilbert spaces, see e.g. [51] and [27].

The case $c_1 = 0$, $0 < c_2 < 1$ is excluded here. It turns out that such condition is too weak to prove convergence rates under the analysis which is presented here. This observation is not surprisingly since no convergence rates results are known even in Hilbert spaces under such condition.

3.1.5 On an example in inverse option pricing theory

We start with the following example which arises in option pricing theory, see e.g. [61]. The corresponding inverse problem was deeply studied in [39], see also the references therein for an overview about further aspects in the mathematical foundation of (inverse) option pricing. We also refer to [45] and [34] for some newer results.

Example 3.1.5 *A European call option on a traded asset is a contract which gives the holder the right to buy the asset at time (maturity) t for a fixed strike price $K > 0$ independent on the actual asset price $X > 0$ at time $t > 0$. For fixed current asset price $X > 0$ and time $t_0 = 0$ we denote with $c(t)$ the (fair) price of such call option with maturity $t \geq 0$. Following the generalization of the classical Black-Scholes analysis with time-dependent volatility function $\sigma(t)$, $t \geq 0$, and constant risk-less short-term interest rate $r \geq 0$ we introduce the Black-Scholes function U_{BS} for the variables $X > 0$, $K > 0$, $r \geq 0$ and $s \geq 0$ as*

$$U_{BS}(X, K, r, t, s) := \begin{cases} X\,\Phi(d_1) - K\,e^{-r\,t}\Phi(d_2), & s > 0, \\ \max\left\{X - K\,e^{-r\,t}, 0\right\}, & s = 0, \end{cases}$$

with

$$d_1 := \frac{\ln\left(\frac{X}{K}\right) + r\,t + \frac{s}{2}}{\sqrt{s}}, \quad d_2 := d_1 - \sqrt{s},$$

and $\Phi(\xi)$, $\xi \in \mathbb{R}$, denotes the cumulative density function of the standard normal distribution. Then the price $c(t)$, $t \in [0, T]$, of the option as function of the maturity t is given *by the formula*

$$c(t) := U_{BS}\left(X, K, r, t, \int_0^t \sigma^2(\tau)\, d\tau\right), \qquad t \in [0, T],$$

where $T > 0$ denotes the maximal time horizon of interest.

This example gives us the motivation to deal with nonlinear operators of following structure. Let $k(t, s)$, $(t, s) \in [a, b] \times \mathbb{R}$ be a continuous function on the domain $[a, b] \times \mathbb{R}$. For given $1 < p, q < \infty$ we set $\mathcal{X} := L^q(a, b)$ and $\mathcal{Y} := L^p(a, b)$. Furthermore, with $\mathcal{D}(F) \subset \mathcal{X}$ we denote the domain of all admissible parameters. Then we can define the nonlinear operator $F : \mathcal{D}(F) \subset \mathcal{X} \longrightarrow \mathcal{Y}$ as

$$[F(x)](t) := k\left(t, [A\, x](t)\right), \qquad t \in [a, b],$$

where $A : \mathcal{X} \longrightarrow L^\infty(a, b)$ denotes an additional bounded linear operator. We state the following lemma.

Lemma 3.1.6 *Let $x_0 \in \mathcal{D}(F)$ be given and $s_0 := A\, x_0$. For given radius $\rho > 0$ let one of the following conditions hold for all $s_0 + \Delta s \in A(\mathcal{D}(F) \cap \mathcal{B}_\rho(x_0))$:*

(i) The function $k(\cdot, \cdot)$ has a continuous derivative with respect to s in a ball around $s_0(t)$, $t \in [a, b]$, and there exist two functions $C_2(t) \le C_1(t)$ such that

$$0 < C_2(t) \le k_s(t, [s_0 + \Delta s](t)) \le C_1(t) < \infty \qquad a.e. \text{ on } \quad [a, b]$$

and $C_\rho := \|(C_1 - C_2)/C_2\|_{L^\infty} < \infty$.

(ii) The function $k(\cdot, \cdot)$ has a Lipschitz-continuous derivative with respect to s in a ball around $s_0(t)$, $t \in [a, b]$, i.e. there exists a function $\tilde{C}_1(t) > 0$ such that

$$|k_s(t, [s_0 + \Delta s](t)) - k_s(t, s_0(t))| \le \tilde{C}_1(t) |\Delta s(t)| \qquad a.e. \text{ on } \quad [a, b]$$

and there exists a function $\tilde{C}_2(t)$ such that

$$\infty > k_s(t, [s_0 + \Delta s](t)) \ge C_2(t) > 0 \qquad a.e. \text{ on } \quad [a, b]$$

and $\tilde{C}_\rho := \|\tilde{C}_1/C_2\|_{L^\infty} < \infty$.

Then there exists a linear bounded operator $F'(x_0)$ and a constant $L_\rho > 0$ such that

$$\|F(x_0 + \Delta x) - F(x_0) - F'(x_0)\, \Delta x\| \le L_\rho \|F(x_0 + \Delta x) - F(x_0)\| \tag{3.10}$$

holds for all Δx with $x_0 + \Delta x \in \mathcal{D}(F) \cap \mathcal{B}_\rho(x_0)$. Moreover, in the second case we have $L_\rho \to 0$ as $\rho \to 0$.

PROOF. For given Δx with $x_0 + \Delta x \in \mathcal{D}(F) \cap \mathcal{B}_\rho(x_0)$ it is a short calculation to see that the (directional) derivative $F'(x_0)\, \Delta x$ is given as

$$[F'(x_0)\, \Delta x](t) := k_s(t, [A\, x_0](t))[A\, \Delta x](t), \qquad t \in [a, b].$$

We set $\Delta s := A\,\Delta x \in L^\infty(a,b)$. In particular we conclude $\|\Delta x\| \le \rho$ and $|\Delta s(t)| \le \|\Delta s\|_{L^\infty} \le \|A\|\,\|\Delta x\| \le \|A\|\,\rho$ a.e. on $[a,b]$.

We start with (i). The mean value theorem gives

$$\begin{aligned}
|[F(x_0+\Delta x)-F(x_0)](t)| &= |[k(t,[s_0+\Delta s](t)) - k(t,s_0(t))|\\
&= \left|\int_0^1 k_s(t,[s_0+\tau\,\Delta s](t))\,\Delta s(t)\,d\tau\right|\\
&\ge C_2(t)|\Delta s(t)|
\end{aligned}$$

a.e. on $[a,b]$. On the other hand we have for some $\tau \in (0,1)$

$$\begin{aligned}
|[F(x_0+\Delta x)-F(x_0)-F'(x_0)\,\Delta x](t)| &= |k_s(t,[s_0+\tau\Delta s](t)) - k_s(t,s_0(t))|\,|\Delta s(t)|\\
&\le (C_1(t)-C_2(t))\,|\Delta s(t)|\\
&\le \frac{C_1(t)-C_2(t)}{C_2(t)}\,|[F(x_0+\Delta x)-F(x_0)](t)|\,.
\end{aligned}$$

This proves the first part with $L_\rho := C_\rho$. The boundedness of $F'(x_0)$ we conclude from the continuity of the function $k(\cdot,\cdot)$ which implies $\|C_1-C_2\|_{L^\infty} \to 0$ as $\rho \to 0$. For (ii) we also derive

$$\begin{aligned}
|[F(x_0+\Delta x)-F(x_0)-F'(x_0)\,\Delta x](t)| &= \left|\int_0^1 (k_s(t,[s_0+\tau\,\Delta s](t)) - k_s(t,s_0(t)))\,\Delta s(t)\,d\tau\right|\\
&\le \tilde{C}_1\frac{|\Delta s(t)|^2}{2}\\
&\le \frac{\tilde{C}_1(t)\|A\|\,\|\Delta x\|}{2\,C_2(t)}\,|[F(x_0+\Delta x)-F(x_0)](t)|
\end{aligned}$$

a.e on $[a,b]$. This proves the boundedness of $F'(x_0)$ as well as the estimate (3.10) with $L_\rho := \frac{\rho}{2}\|A\|\|\tilde{C}_1/C_2\|_{L^\infty}$. ■

We return to our example. It turns out that we cannot apply Lemma 3.1.6 directly to the option pricing problem. We therefore present two different approaches leading to different degrees of nonlinearity.

Example 3.1.6 (First variant) *We set $[a,b] := [0,T]$ and define the nonlinear operator $F : \mathcal{D}(F) \subset L^q(0,T) \longrightarrow L^p(0,T)$, $1 < p,q < \infty$, as*

$$[F(x)](t) := U_{BS}\left(X,K,r,t,\int_0^t x(\tau)\,d\tau\right), \qquad t \in [0,T],$$

with domain $\mathcal{D}(F) := \{x \in \mathcal{X} \;:\; x(t) \ge \underline{c}$ a.e. on $[0,T]\}$. We further use the notation $k(t,s) := U_{BS}(X,K,r,t,s)$ and assume $x_0 \in \mathcal{D}(F)$. Moreover, $A : L^q(0,T) \longrightarrow L^\infty(0,T)$ is the operator of integration which was already introduced in Example 1.2.1. Then we have for $s_0 := A\,x_0$ that $s_0(t) \ge \underline{c}\,t$, $t \in [0,T]$. Moreover we obtain

$$|A\,x(t)| = \left|\int_0^t x(\tau)\,d\tau\right| \le \|1\|_{L^{q/(q-1)}(0,t)}\|x\|_{L^q} = t^{\frac{q-1}{q}}\,\|x\|_{L^q}.$$

Hence for $x_0 + \Delta x \in \mathcal{D}(F) \cap \mathcal{B}_\rho(x_0)$ *we observe with* $\Delta s := A\,\Delta x$ *that* $|\Delta s(t)| \le \rho\, t^{\frac{q-1}{q}}$, $t \ge 0$, *and*

$$\tilde{c}(t) := \max\left\{\underline{c}\,t, s_0(t) - \rho\, t^{\frac{q-1}{q}}\right\} \le [s_0 + \Delta s](t) \le s_0(t) + \rho\, t^{\frac{q}{q-1}}, \qquad t \in [0,T].$$

Furthermore we see with $L_X := \ln\left(\frac{X}{K}\right)$ *that*

$$k_s(t,s) := \frac{X}{2\sqrt{2\,\pi\, s}} \exp\left(-\frac{(L_X + r\,t)^2}{2\,s} - \frac{L_X + r\,t}{2} - \frac{s}{8}\right) > 0$$

We can estimate

$$\begin{aligned} k_s(t, [s_0 + \Delta s](t)) \;&\le\; \frac{X}{2\sqrt{2\,\pi\,\tilde{c}(t)}} \exp\left(-\frac{(L_X + r\,t)^2}{2\,(s_0(t) + \rho\, t^{\frac{q-1}{q}})} - \frac{L_X + r\,t}{2} - \frac{s_0(t) - \rho\, t^{\frac{q-1}{q}}}{8}\right) \\ &=: \; C_1(t) \end{aligned}$$

and

$$\begin{aligned} k_s(t, [s_0 + \Delta s](t)) \;&\ge\; \frac{X \exp\left(-\frac{(L_X + r\,t)^2}{2\,\tilde{c}(t)} - \frac{L_X + r\,t}{2} - \frac{s_0(t) + \rho\, t^{\frac{q-1}{q}}}{8}\right)}{2\sqrt{2\,\pi (s_0(t) + \rho\, t^{\frac{q-1}{q}})}} \\ &=: \; C_2(t). \end{aligned}$$

It is a short observation to see that for $t \to 0$ *we have* $C_i(t) \to 0$ *if* $L_X \ne 0$ *and* $C_i(t) \to \infty$ *if* $L_X = 0$, $i = 1, 2$. *Therefore we consider the quotient* $C_1(t)/C_2(t)$. *Here we obtain*

$$\begin{aligned} \frac{C_1(t)}{C_2(t)} &= \sqrt{\frac{s_0(t) + \rho\, t^{\frac{q-1}{q}}}{\tilde{c}(t)}} \exp\left(-\frac{(L_X + r\,t)^2}{2}\left(\frac{1}{s_0(t) + \rho\, t^{\frac{q-1}{q}}} - \frac{1}{\tilde{c}(t)}\right) + \frac{\rho\, t^{\frac{q-1}{q}}}{4}\right) \\ &= \sqrt{\frac{s_0(t) + \rho\, t^{\frac{q-1}{q}}}{\tilde{c}(t)}} \exp\left(\frac{(L_X + r\,t)^2}{2} \frac{s_0(t) + \rho\, t^{\frac{q-1}{q}} - \tilde{c}(t)}{\tilde{c}(t)(s_0(t) + \rho\, t^{\frac{q-1}{q}})} + \frac{\rho\, t^{\frac{q-1}{q}}}{4}\right). \end{aligned}$$

Moreover, we have

$$1 \le \frac{s_0(t) + \rho\, t^{\frac{q-1}{q}}}{\tilde{c}(t)} = 1 + \frac{s_0(t) + \rho\, t^{\frac{q-1}{q}} - \tilde{c}(t)}{\tilde{c}(t)} \le 1 + \frac{2\,\rho\, t^{\frac{q-1}{q}}}{\underline{c}\,t} = 1 + \frac{2\,\rho}{\underline{c}} t^{-\frac{1}{q}}$$

since $\tilde{c}(t) \ge \underline{c}\,t$ *and* $\tilde{c}(t) \ge s_0(t) - \rho\, t^{\frac{q-1}{q}}$. *Furthermore we estimate*

$$I_1 := \frac{s_0(t) + \rho\, t^{\frac{q-1}{q}} - \tilde{c}(t)}{\tilde{c}(t)(s_0(t) + \rho\, t^{\frac{q-1}{q}})} \le \frac{2\,\rho\, t^{\frac{q-1}{q}}}{\underline{c}\,t\,(s_0(t) + \rho\, t^{\frac{q-1}{q}})} \le \frac{2\,\rho\, t^{\frac{q-1}{q}}}{\underline{c}^2 t^2} = \frac{2\,\rho}{\underline{c}^2} t^{-\frac{q+1}{q}}$$

since $\tilde{c}(t) \ge s_0(t) - \rho\, t^{\frac{q-1}{q}}$, $\tilde{c}(t) \ge \underline{c}\,t$ *and* $s_0(t) + \rho\, t^{\frac{q-1}{q}} \ge s_0(t) \ge \underline{c}\,t$. *On the other hand we can derive*

$$I_1 \ge \frac{\rho\, t^{\frac{q-1}{q}}}{s_0(t)(s_0(t) + \rho\, t^{\frac{q-1}{q}})} \ge \frac{\rho}{\|x_0\|(\|x_0\| + \rho)} t^{-\frac{q-1}{q}}$$

where we used that $\tilde{c}(t) \le s_0(t) \le \|x_0\|\, t^{\frac{q-1}{q}}$ *and* $s_0(t) + \rho\, t^{\frac{q-1}{q}} \le (\|x_0\| + \rho)\, t^{\frac{q-1}{q}}$. *This gives finally*

$$
\begin{aligned}
1 &\le \exp\left[\rho\left(\frac{(L_X + r\,t)^2}{2\,\|x_0\|(\|x_0\|+\rho)} t^{-\frac{q+1}{q}} + \frac{t^{\frac{q-1}{q}}}{4}\right)\right] \\
&\le \frac{C_1(t)}{C_2(t)} \le \sqrt{1 + \frac{2\,\rho}{\underline{c}} t^{-\frac{1}{q}}}\, \exp\left[\rho\left(\frac{(L_X + r\,t)^2}{\underline{c}^2} t^{-\frac{q+1}{q}} + \frac{t^{\frac{q-1}{q}}}{4}\right)\right]
\end{aligned}
$$

Now we have to distinguish between $L_X = 0$ *and* $L_X \neq 0$. *For* $L_X \neq 0$ *both sides of the inequality grows exponentially to infinity as* $t \to 0$. *Hence we can cannot bound the quotient* $C_1(t)/C_2(t)$ *from above. For* $L_X = 0$ *we conclude*

$$
(L_X + r\,t)^2 t^{-\frac{q+1}{q}} = r^2 t^{\frac{q-1}{q}} \to 0 \quad as \quad t \to 0.
$$

The estimate above reduces to

$$
0 \le \frac{C_1(t) - C_2(t)}{C_2(t)} \le \left(\sqrt{T^{-\frac{1}{q}} + \frac{2\,\rho}{\underline{c}}}\, \exp\left[\rho\, T^{\frac{q-1}{q}}\left(\frac{r^2}{\underline{c}^2} + \frac{1}{4}\right)\right] - T^{-\frac{1}{2q}}\right) t^{-\frac{1}{2q}}.
$$

Additionally we observe

$$
C_2(t) \le \frac{X}{2\sqrt{2\,\pi\,\rho}} t^{-\frac{q-1}{2q}}, \qquad t \in (0, T].
$$

This leads with $T(x_0) := F(x_0 + \Delta x) - F(x_0) - F'(x_0)\,\Delta x$ *and* $0 < \nu < 1$ *to*

$$
\begin{aligned}
|[T(x_0)](t)| &\le |(C_1(t) - C_2(t))\,\Delta s(t)| \\
&= |C_1(t) - C_2(t)|\,|\Delta s(t)|^{\nu} |\Delta s(t)|^{1-\nu} \\
&\le |C_1(t) - C_2(t)|\,|\Delta s(t)|^{\nu} \left|\frac{[F(x_0 + \Delta x) - F(x_0)](t)}{C_2(t)}\right|^{1-\nu} \\
&\le \frac{C_1(t) - C_2(t)}{C_2(t)}\,|\Delta s(t)|^{\nu} C_2(t)^{\nu}\,|[F(x_0 + \Delta x) - F(x_0)](t)|^{1-\nu} \\
&\le C\,t^{\frac{\nu(q-1)-1}{2q}}\,|[F(x_0 + \Delta x) - F(x_0)](t)|^{1-\nu}
\end{aligned}
$$

for some constant $C > 0$. *We want to apply a variant of Hölder's inequality. Let be* $1 < \mu_1, \mu_2 < \infty$, $\kappa > 0$ *be given and* $x \in L^{\mu_1}(a,b)$, $y \in L^{\mu_2}(a,b)$. *If there exists* $1 < \mu_3 < \infty$ *such that* $\mu_3^{-1} = \mu_1^{-1} + (\kappa/\mu_2)$ *then* $x\,y^{\kappa} \in L^{\mu_3}(a,b)$ *and*

$$
\|x\,y^{\kappa}\|_{L^{\mu_3}} \le \|x\|_{L^{\mu_1}} \|y\|_{L^{\mu_2}}^{\kappa}
$$

holds. We set $\kappa := 1 - \nu$ *and* $\mu_2 = \mu_3 = p$. *Then with* $r := \mu_1 = \frac{p}{1-\kappa} = \frac{p}{\nu}$ *and* $f(t) := t^{\frac{\nu(q-1)-1}{2q}}$ *we have* $f \in L^r(0,T)$ *if and only if*

$$
r\,\frac{\nu(q-1)-1}{2\,q} = \frac{p(\nu(q-1)-1)}{2\,q\,\nu} > -1 \;\Leftrightarrow\; \nu < \frac{p}{p(q-1) + 2\,q} < 1
$$

In order to ensure the latter inequality we observe

$$\frac{p}{p(q-1)+2q} < 1 \Leftrightarrow p(2-q) < 2q.$$

This is fulfilled for all $p > 1$ *if* $q \geq 2$ *and for* $1 < p < \frac{2q}{2-q}$ *if* $1 < q < 2$. *Then we obtain*

$$\|T(x_0)\|_{L^p} \leq C\,\|f\|_{L^r}\|F(x_0+\Delta x) - F(x_0)\|_{L^p}^{\kappa}.$$

In particular we derived a nonlinearity of degree $(\kappa, 0)$ *with* $0 < \kappa < 1$.

Excluding the case of maturities t close to zero we can show an alternative result.

Example 3.1.7 (Second variant) *In order to apply the second result of the above lemma in this situation we introduce a (small) constant* $t_\varepsilon > 0$ *and assume the volatility to be known (and constant) on the interval* $[0, t_\varepsilon]$, *i.e.* $\sigma(t) \equiv \sigma_0$, $t \in [0, t_\varepsilon]$. *Then we set* $[a, b] := [t_\varepsilon, T]$ *and define the nonlinear operator* $F : \mathcal{D}(F) \subset L^q(t_\varepsilon, T) \longrightarrow L^p(t_\varepsilon, T)$ *as*

$$[F(x)](t) := U_{BS}\left(X, K, r, t, \sigma_0^2 t_\varepsilon + \int_{t_\varepsilon}^{t} x(\tau)\, d\tau\right), \qquad t \in [t_\varepsilon, T],$$

with domain $\mathcal{D}(F) := \{x \in \mathcal{X} \,:\, x(t) \geq \underline{c} \;\; a.e. \text{ on } [t_\varepsilon, T]\}$. *We have*

$$k_{ss}(t,s) := -\frac{X}{4\sqrt{2\pi}\, s}\left(-\frac{(L_X + r\,t)^2}{s^2} + \frac{1}{4} + \frac{1}{s}\right)\exp\left(-\frac{(L_X + r\,t)^2}{2\,s} - \frac{L_X + r\,t}{2} - \frac{s}{8}\right).$$

We estimate

$$\begin{aligned} |k_{ss}(t,[s_0+\Delta s](t))| \;&\leq\; \frac{X}{4\sqrt{2\pi\,\tilde{c}(t)}}\left(\frac{(L_X + r\,t)^2}{\tilde{c}(t)^2} + \frac{1}{4} + \frac{1}{\tilde{c}(t)}\right) \\ &\qquad \cdot \exp\left(-\frac{(L_X + r\,t)^2}{2\,(s_0(t) + \rho\, t^{\frac{q-1}{q}})} - \frac{L_X + r\,t}{2} - \frac{s_0(t) - \rho\, t^{\frac{q-1}{q}}}{8}\right) \\ &=: \; \tilde{C}_1(t). \end{aligned}$$

The lower bound $C_2(t)$ *on the derivative* $k_s(t, [s_0 + \Delta s])(t)$ *we can take from the previous example. Hence we obtain*

$$\frac{\tilde{C}_1(t)}{C_2(t)} = \tilde{I}_2 \exp\left(\frac{(L_X + r\,t)^2}{2} I_1 + \frac{\rho\, t^{\frac{q-1}{q}}}{4}\right)$$

with I_1 *as in the previous example and*

$$\tilde{I}_2 := \frac{1}{2}\sqrt{\frac{s_0(t) + \rho\, t^{\frac{q-1}{q}}}{\tilde{c}(t)}}\left(\frac{(L_X + r\,t)^2}{\tilde{c}(t)^2} + \frac{1}{4} + \frac{1}{\tilde{c}(t)}\right).$$

Then we can apply Lemma 3.1.6(ii) by observing that

$$\|\tilde{C}_1/C_2\|_{L^\infty} = \max_{t \in [t_\varepsilon, T]} \tilde{I}_2 \exp\left(\frac{(L_X + r\,t)^2}{2} I_1 + \frac{\rho\, t^{\frac{q-1}{q}}}{4}\right) < \infty.$$

We now additionally assume that we have chosen the penalty functional $P(x) := \frac{1}{s}\|x\|^s$ *with* $s := \max\{q, 2\}$. *Then we obtain from Lemma 3.1.6(ii) and the* s*-coercivity of the functional* P*, that with* $\xi_0 := J_r(x_0)$ *we have*

$$\begin{aligned}\|F(x_0 + \Delta x) - F(x_0) - F'(x_0)\,\Delta x\|_{L^p} &\le C\,\|\Delta x\|_{L^q}\|F(x_0 + \Delta x) - F(x_0)\|_{L^p} \\ &\le C\left(\frac{s}{C_s}\right)^{\frac{1}{s}} D_{\xi_0}(x_0 + \Delta x, x_0)^{\frac{1}{s}}\|F(x_0 + \Delta x) - F(x_0)\|_{L^p},\end{aligned}$$

i.e. we derived a degree of nonlinearity of $(1, \max\{q, 2\}^{-1})$.

3.1.6 The principle of a-priori balancing

The derivation of convergence rates in the following always bases on the same principle. Starting point is an inequality of the form

$$D \le C_1 f_1(\delta, \alpha) + C_2 f_2(\alpha, R) + C_3 f_3(R)$$

which holds for all $\delta \in [0, \delta_{max}]$, $\alpha \in (0, \alpha_{max}]$ and $R \in [0, \infty)$. Here, C_1, C_2 and C_3 are positive constants whereas the functions $f_1 : [0, \delta_{max}] \times (0, \alpha_{max}] \longrightarrow [0, \infty)$, $f_2 : (0, \alpha_{max}] \times [0, \infty) \longrightarrow [0, \infty)$ and $f_3 : [0, \infty) \longrightarrow [0, \infty)$ fulfill the following assumptions:

(i) For all $\alpha > 0$ the function f_1 is an index function with respect to the variable δ. For fixed $\delta > 0$ the function is strictly decreasing with respect to α.

(ii) The function f_2 is an index function with respect to α and R keeping the other variable positive but fixed. Moreover we have $f_2(\alpha, R) \to \infty$ as $R \to \infty$ for all $\alpha > 0$.

(iii) The function f_3 is a nonnegative decreasing function with $f_3(R) \to 0$ as $R \to \infty$.

This conditions implies, that the functions introduced in the following are well-defined. Now we can present the balancing procedure which we will apply several times in the further convergence rates analysis of the monograph.

In the first step we derive a function $\Theta : [R_{min}, \infty) \longrightarrow (0, \alpha_{max}]$ such that $f_2(\alpha_{max}, R_{min}) = f_3(R_{min})$ and

$$f_2(\Theta(R), R) = f_3(R)$$

holds for all $R \ge R_{min}$. By the condition above, Θ is a strictly decreasing function with $\Theta(R) \to 0$ as $R \to \infty$. Afterwards we find a second function $\Phi : [R_{min}, \infty) \longrightarrow [0, \delta_{max}]$ (δ_{max} chosen sufficiently large or - alternatively - α_{max} is sufficiently small) such that

$$f_1(\Phi(R), \Theta(R)) = f_3(R), \qquad \forall\, R \ge R_{min}.$$

Again, Φ is strictly decreasing with $\Phi(R) \to 0$ as $R \to \infty$. In particular both functions are invertible. Then we arrive at the inequality

$$D \le (C_1 + C_2 + C_3)\, f_3\left(\Phi^{-1}(\delta)\right), \quad \delta \in [0, \delta_{max}],$$

i.e. the function $f_3(\Phi^{-1}(\delta))$ describes the corresponding convergence rate. Moreover, we introduce a third function $\Psi : (0, \alpha_{max}] \longrightarrow [0, \delta_{min}]$ which satisfies

$$f_1(\Psi(\alpha), \alpha) = f_3(\Theta^{-1}(\alpha)), \qquad \forall\, 0 < \alpha \leq \alpha_{max}.$$

It is a short calculation to see that Ψ an index function. Then $\alpha := \Psi^{-1}(\delta)$ describes the corresponding choice of the regularization parameter α.

In the following error analysis we pursue the following strategy: first we derive error bounds of the above structure. In a second step we apply the described balancing technique for proving the convergence rates results based on a corresponding proper a-priori choice of the regularization parameter α.

3.2 Low-order convergence rates

3.2.1 Approximate source conditions and distance functions

For deriving error bounds and convergence rates the element $\xi^\dagger \in \mathcal{X}^*$ of (A6) has to fulfill an additional smoothness condition. Typically, a source condition

$$\xi^\dagger = F'(x^\dagger)^\star \omega, \qquad \omega \in \mathcal{Y}^*,\ \|\omega\| \leq R, \tag{3.11}$$

for some $R \geq 0$ is supposed, see e.g. [12], [81] and [82]. Here, we follow a more general strategy. In particular, we are also interested in convergence rates results when this source condition is violated for all $R > 0$, i.e. $\xi^\dagger \notin \mathcal{R}(F'(x^\dagger)^\star)$ holds. Therefore we refer to (3.11) as reference or benchmark source condition and weaken this assumption by assuming a so-called approximative source condition

$$\xi^\dagger = F'(x^\dagger)^\star \omega + v, \qquad \omega \in \mathcal{Y}^*,\ v \in \mathcal{X}^*,\ \|\omega\| \leq R,\ \|v\| \leq d. \tag{3.12}$$

for some $R, d \geq 0$. Of course, the representation elements ω and v of the approximative source condition (3.12) are not unique for given $\xi^\dagger \in \mathcal{X}^*$. On the other hand, exactly this observation will be used in the next sections for deriving error bounds and convergence rates. For presenting a unified framework we introduce the sets

$$\mathcal{M}_L(R, d) := \left\{\xi \in \mathcal{X}^* \ :\ \xi = F'(x^\dagger)^\star \omega + v,\ \omega \in \mathcal{Y}^*,\ v \in \mathcal{X}^*,\ \|\omega\| \leq R,\ \|v\| \leq d\right\}.$$

for each $R, d \geq 0$. The interpretation of these sets is quite simple: for given $\xi \in \mathcal{X}^*$ the number $d \geq 0$ describes the maximal violation of the reference source condition (3.11) which is allowed, when the norm of the source element $\omega \in \mathcal{Y}^*$ is bounded by some constant $R > 0$. Clearly, $\xi^\dagger \in \mathcal{M}_L(R, 0)$ is equivalent to the availability of the source condition (3.11).

For presenting convergence rates in terms of approximate source conditions we apply the idea of distance functions, see e.g. [7, Theorem 6.8].

Definition 3.2.1 *For given $\xi \in \mathcal{X}^*$ the distance function $d(\cdot;\xi) : [0,\infty) \longrightarrow \mathbb{R}$ is defined as*

$$d(R;\xi) := \inf \left\{ \|\xi - F'(x^\dagger)^\star \omega\| \; : \; \omega \in \mathcal{Y}^*, \; \|\omega\| \le R \right\}, \qquad R \ge 0.$$

Note, that the nonnegative function $d(R,\xi)$ is well-defined for each $\xi \in \mathcal{X}^*$. Moreover, the following properties obviously hold:

(i) For each $R \ge 0$ there exists an element $\omega = \omega(R)$ with $d(R;\xi) = \|\xi - F'(x^\dagger)^\star \omega\|$ and $\|\omega\| \le R$. This is an immediate consequence of the Alaoglu-Bourbaki theorem which says that the unit ball in $\mathcal{Y}^*$ is weakly* compact, see e.g. [94, Theorem VIII.3.11].

(ii) The distance function $d(R;\xi)$ is non-increasing.

(iii) If $\xi \in \overline{\mathcal{R}(F'(x^\dagger)^\star)}$ the we have $d(R;\xi) \to 0$ for $R \to \infty$.

(iv) If $\xi \notin \mathcal{R}(F'(x^\dagger)^\star)$ then $d(R;\xi) > 0$ holds for all $R \ge 0$.

(v) If $\xi = F'(x^\dagger)^\star \omega$ and $\|\omega\| = R$ then $d(R;\xi) = 0$ for all $R \ge \|\omega\|$ is valid.

Altogether, the distance functions $d(R;\xi)$ gives us a quantity for measuring the violation of the source condition (3.11).

We now are able to present convergence rates, if the source condition (3.11) is not satisfied, but $\xi^\dagger \in \overline{\mathcal{R}(F'(x^\dagger)^\star)}$. This additional assumption seems to be very restrictive at first view. On the other hand, if $\mathcal{X}$ is additionally supposed to be reflexive then we can find the following:

- If $F'(x^\dagger)$ is injective, then $\overline{\mathcal{R}(F'(x^\dagger)^\star)} = \mathcal{X}^*$, see e.g. [94, Satz III.4.5].
- For linear operators A and $P(x) = \|x\|^q/q$ for some $q > 1$ it can be shown that the P-minimum solution $x^\dagger$ of $Ax = y$ is unique and $P'(x^\dagger) = J_q(x^\dagger) \in \overline{\mathcal{R}(A^\star)}$, see [88, Lemma 2.10].

Before we continue with our theoretical considerations we present a small example where we derive at least an upper bound for the distance function in a specific situation, see also [36].

Example 3.2.1 *We consider the operator A of Example 1.2.1 for arbitrary $1 < p, q < \infty$. The dual operator $A^\star : L^{p^*}(0,1) \longrightarrow L^{q^*}(0,1)$ given as*

$$[A^\star y](t) := \int_t^1 y(\tau)\, d\tau, \quad t \in [0,1],$$

$y \in L^{p^*}(0,1)$. With p^* and q^* we again denote the dual exponents of p and q. We choose $x^\dagger \equiv 1$. Then $\xi^\dagger = J_q(x^\dagger) \equiv 1$ and obviously $\xi^\dagger \notin \mathcal{R}(A^\star)$ since the boundary condition $\xi^\dagger(1) = 0$ is violated. However, we can find an upper bound for the distance function $d(R;\xi^\dagger)$, $R \ge 0$. Therefore we define the functions

$$\omega_R(t) := \begin{cases} 0, & 0 \le t < 1 - R^{-p}, \\ R^p, & 1 \ge t \ge 1 - R^{-p}, \end{cases}$$

for $R \ge 1$. Then $\|\omega_R\|_{L^{p^*}} = \left(R^{-p} R^{p^* p}\right)^{\frac{1}{p^*}} = R$ since $p(p^* - 1) = p(\frac{p}{p-1} - 1) = p\frac{1}{p-1} = p^*$.

Moreover, we have

$$(A^\star\omega_R)(t) := \begin{cases} 1, & 0 \le t < 1 - R^{-p}, \\ R^p(1-t), & 1 \ge t \ge 1 - R^{-p}, \end{cases}$$

and hence

$$\begin{aligned} \|J_q(x^\dagger) - A^\star\omega_R\|_{L^{q^*}} &= \left[\int_0^{R^{-p}} (R^p t)^{q^*}\, dt\right]^{\frac{1}{q^*}} \\ &= R^p \left(\frac{1}{q^*+1}\right)^{\frac{1}{q^*}} R^{-p\frac{q^*+1}{q^*}} = \left(\frac{1}{q^*+1}\right)^{\frac{1}{q^*}} R^{\frac{1-q}{q}p}. \end{aligned}$$

We conclude for the distance function of $\xi^\dagger$*, that* $d(R;\xi^\dagger) \le R^{\frac{1-q}{q}p}/(q^*+1)^{\frac{1}{q^*}}$, $R \ge 1$. *In particular we derived a upper bound for the distance function.*

This idea of deriving bounds for the distance functions can be generalized to arbitrary linear operators. We also remark the following:

- We can still derive error bounds and convergence rates when we replace the distance function in the error bound and convergence rate results in the next subsections by any upper bound. The consequences are suboptimal error estimates and possibly lower convergence rates.
- The approximation was done in this example by piecewise constants functions ω_R. A (more complicated) calculation with piecewise linear (or polynomial) ansatz functions leads to the same rate of the distance functions. So the order of the estimated bound seems to be correct. This was already mentioned in Example 2.4.3 in Section 2.4 for the case $p = q = 2$.

3.2.2 Error bounds

We are now able to derive our first error bound results. Since they are of interest on its own they are presented here separately. Later on, we use these estimates for proving the corresponding convergence rates which is essentially a balancing of all summands of the error bounds. Basic tools of the subsequent analysis are Young's inequality

$$a\,b \le \frac{a^{p_1}}{p_1} + \frac{b^{p_2}}{p_2}, \qquad a, b > 0, \tag{3.13}$$

in its classical form and the modification (Young's inequality with ε)

$$a\,b \le \varepsilon\, a^{p_1} + \frac{b^{p_2}}{(\varepsilon p_1)^{(p2/p1)} p_2}, \qquad a, b, \varepsilon > 0, \tag{3.14}$$

for dual exponents $p_1, p_2 > 1$.

First, we reformulate the approximate source condition (3.12) as variational inequality, see [40, Lemma 3.1].

Lemma 3.2.1 *Assume F to be of degree (c_1, c_2) and $\xi^\dagger \in \mathcal{M}_L(R,d)$ for some $R, d \geq 0$. Then the variational inequality*

$$|\langle \xi^\dagger, x - x^\dagger \rangle| \leq \beta_1(R)\, D_{\xi^\dagger}(x, x^\dagger) + \beta_2(R)\, \|F(x) - F(x^\dagger)\|^\kappa + d\|x - x^\dagger\| \tag{3.15}$$

holds for all $x \in \mathcal{S}_{\alpha,y}(\varrho_{max})$, where

(i) $\kappa = \frac{c_1}{1-c_2}$, $\beta_1(R) \equiv c_2$ and $\beta_2(R) := (1-c_2)(L\,R)^{\frac{1}{1-c_2}} + (p\varrho_{max})^{(1-\kappa)/p} R$ for $c_1 > 0$ and

(ii) $\kappa = 1$, $\beta_1(R) := L\,R$ and $\beta_2(R) := R$ in the case $c_2 = 1, c_1 = 0$.

PROOF. By assumption, we have $\xi^\dagger = F'(x^\dagger)^\star \omega + v$ with $\|\omega\| \leq R$ and $\|v\| \leq d$. Then

$$\begin{aligned} |\langle \xi^\dagger, x - x^\dagger \rangle| &= |\langle F'(x^\dagger)^\star \omega + v, x - x^\dagger \rangle| \\ &\leq \|\omega\| \|F'(x^\dagger)(x - x^\dagger)\| + \|v\| \|x - x^\dagger\| \\ &\leq L\|\omega\| \|F(x) - F(x^\dagger)\|^{c_1} D_{\xi^\dagger}(x, x^\dagger)^{c_2} + \|\omega\| \|F(x) - F(x^\dagger)\| \\ &\quad + \|v\| \|x - x^\dagger\|. \end{aligned}$$

Assume $c_1, c_2 > 0$. We apply (3.13) with $p_1 = c_2^{-1}$, $p_2 = 1/(1-c_2)$, $a = D_{\xi^\dagger}(x, x^\dagger)$ and $b = L\|\omega\| \|F(x) - F(x^\dagger)\|^{c_1}$ to derive

$$\begin{aligned} |\langle \xi^\dagger, x - x^\dagger \rangle| &\leq c_2 D_{\xi^\dagger}(x, x^\dagger) + (1-c_2) L^{\frac{1}{1-c_2}} \|\omega\|^{\frac{1}{1-c_2}} \|F(x) - F(x^\dagger)\|^{\frac{c_1}{1-c_2}} \\ &\quad + \|\omega\| \|F(x) - F(x^\dagger)\| + \|v\| \|x - x^\dagger\|. \end{aligned}$$

We remark that the same estimate holds also for $c_2 = 0$. From the definition of the set $\mathcal{S}_{\alpha,y}(\varrho_{max})$ we have $\|F(x) - F(x^\dagger)\| \leq (p\,\varrho_{max})^{1/p}$ and hence

$$\|\omega\| \|F(x) - F(x^\dagger)\| \leq \|\omega\| \|F(x) - F(x^\dagger)\|^\kappa (p\,\varrho_{max})^{\frac{1-\kappa}{p}}.$$

Applying the bounds on $\|\omega\|$ and $\|v\|$ we have shown the first part. The second part $c_1 = 0$, $c_2 = 1$ follows immediately. ■

It has been established in [45] to rewrite the (approximate) source condition as variational inequality. In fact, the proof of the subsequent error bounds (and corresponding convergence rates) are essentially based on the inequality (3.15). The approximate source condition (3.12) can be considered as a sufficient condition for the validity of (3.15).

Remark 3.2.1 *In [87, Chapter 3] further discussions about the relation between source conditions and variational inequalities of the form (3.15) can be found. If, in particular (3.15) holds for some $R > 0$ with $\kappa = 1$ and $d = 0$ then [87, Proposition 3.38] implies the validity of the source condition (3.11) for some $\omega \in \mathcal{Y}^*$. In [43] we find a result that an exponent $\kappa > 1$ in inequality (3.15) requires $\xi^\dagger = 0$.*

Remark 3.2.2 *For linear operators A the estimate (3.15) simplifies to*

$$|\langle \xi^\dagger, x - x^\dagger \rangle| \leq R\|A\,x - A\,x^\dagger\| + d\,\|x - x^\dagger\|,$$

i.e. $\beta_1 = 0$, $\beta_2 = R$ *and* $\kappa = 1$ *holds. This setting is included in Lemma 3.2.1 with* $L = 0$, $c_1 = 1$ *and* $c_2 = 0$. *It turns also out, that we cannot improve the constants in the variational inequality (3.15) by direct applying the linearity of the operator. So, in this section, we deal with linear equations as special case of the (in general) nonlinear problems with this specific choice of the underlying parameters.*

We are now able to present a first error bound result. In the following we assume $p > 1$ since the case $p = 1$ needs some special treatment.

Lemma 3.2.2 *Assume (A1)-(A6) and* $\xi^\dagger \in \mathcal{M}_L(R, d)$ *for some* $R, d \geq 0$. *Then the estimate*

$$D_{\xi^\dagger}(x_\alpha^\delta, x^\dagger) \leq \frac{\frac{2}{p}\delta^p + \frac{2(p-\kappa)\kappa^{\kappa/(p-\kappa)}}{p}(\alpha\,\beta_2)^{p/(p-\kappa)} + \varrho_x\,\alpha\,d}{\alpha(1-\beta_1)} \tag{3.16}$$

holds whenever $\beta_1 < 1$. *Moreover, if* $c_1 > 0$, $R \geq R_0$ *for arbitrary chosen* $R_0 > 0$, *then there exist constants* $K_1, K_2, K_3 > 0$ *– independent on* δ, α, R *and* d *– such that*

$$D_{\xi^\dagger}(x_\alpha^\delta, x^\dagger) \leq K_1\frac{\delta^p}{\alpha} + K_2\alpha^{\frac{\kappa}{p-\kappa}}R^{\frac{p}{(1-c_2)(p-\kappa)}} + K_3 d. \tag{3.17}$$

Proof. From the minimizing property of x_α^δ we conclude $T_{\alpha,y^\delta}(x_\alpha^\delta) \leq T_{\alpha,y^\delta}(x^\dagger)$. Using Lemma 3.2.1 we derive

$$\begin{aligned}
&\frac{1}{p}\|F(x_\alpha^\delta) - y^\delta\|^p + \alpha D_{\xi^\dagger}(x_\alpha^\delta, x^\dagger)\\
&\quad\leq \frac{1}{p}\|F(x^\dagger) - y^\delta\|^p + \alpha\left(P(x^\dagger) - P(x_\alpha^\delta) + D_{\xi^\dagger}(x_\alpha^\delta, x^\dagger)\right)\\
&\quad\leq \frac{\delta^p}{p} + \alpha\langle\xi^\dagger, x^\dagger - x_\alpha^\delta\rangle\\
&\quad\leq \frac{\delta^p}{p} + \alpha\left(\beta_1 D_{\xi^\dagger}(x_\alpha^\delta, x^\dagger) + \beta_2\|F(x_\alpha^\delta) - F(x^\dagger)\|^\kappa + d\,\|x_\alpha^\delta - x^\dagger\|\right)\\
&\quad\leq \frac{\delta^p}{p} + \alpha\left(\beta_1 D_{\xi^\dagger}(x_\alpha^\delta, x^\dagger) + \beta_2\left(\|F(x_\alpha^\delta) - y^\delta\|^\kappa + \delta^\kappa\right) + d\,\|x_\alpha^\delta - x^\dagger\|\right)
\end{aligned}$$

by applying the inequality $(a+b)^\kappa \leq a^\kappa + b^\kappa$ for $a, b \geq 0$ and $0 < \kappa \leq 1$ in the last step. From Lemma 3.1.1 we conclude $\|x_\alpha^\delta - x^\dagger\| \leq \varrho_x$. Furthermore, we apply (3.14) twice with $\varepsilon = \frac{1}{p}$, $p_1 = \frac{p}{\kappa}$, $p_2 = \frac{p}{p-\kappa}$, $b = \alpha\beta_2$ and $a = \|F(x_\alpha^\delta) - y^\delta\|^\kappa$ on the one and $a = \delta^\kappa$ on the other hand. Then

$$\alpha\left(\beta_2\|F(x_\alpha^\delta) - y^\delta\|^\kappa + \delta^\kappa\right) \leq \frac{1}{p}\|F(x_\alpha^\delta) - y^\delta\|^p + \frac{\delta^p}{p} + 2\kappa^{\frac{\kappa}{p-\kappa}}\frac{p-\kappa}{\kappa}(\alpha\beta_2)^{\frac{p}{p-\kappa}}$$

holds with proves the estimate (3.16).

Let us now assume $c_1 > 1$. Since $\|F(x_\alpha^\delta) - y^\delta\| \leq \varrho_y$ we can replace the constant ϱ_{max} by (the smaller constant) ϱ_y in the definition of β_2. Furthermore, for $R \geq R_0$ we conclude

$$\beta_2 = (1-c_2)(L\,R)^{\frac{1}{1-c_2}} + (p\,\varrho_y)^{\frac{1-\kappa}{p}}\,R \leq \left((1-c_2)L^{\frac{1}{1-c_2}} + (p\,\varrho_y)^{\frac{1-\kappa}{p}}\,R_0^{-\frac{c_2}{1-c_2}}\right)R^{\frac{1}{1-c_2}}.$$

Hence (3.17) holds with

$$K_1 := \frac{2}{p(1-c_2)}, \quad K_2 := L^{\frac{1}{1-c_2}} + \frac{(p\,\varrho_y)^{\frac{1-\kappa}{p}}}{(1-c_2)R_0^{\frac{c_2}{1-c_2}}} \quad \text{and} \quad K_3 := \frac{\varrho_x}{1-c_2}.$$

The proof is complete. ■

Remark 3.2.3 *In the case $c_1 = 1$, $c_2 = 0$ the constants simplifies to*

$$K_1 = \frac{2}{p}, \quad K_2 = 1 + L \quad \textit{and} \quad K_3 = \varrho_x.$$

In particular, $K_2 = 1$ holds for linear equations. This is the result in [32, Lemma 7.1].

We now establish a connection between the error bound (3.17) and distance functions. Assume $\xi^\dagger$ has distance function $d(R) = d(R; \xi^\dagger)$, then $\xi^\dagger \in \mathcal{M}_L(R, d(R))$ is valid for all $R \geq 0$. In particular, the following holds.

Corollary 3.2.1 *Assume (A1)-(A6) and $\xi^\dagger$ has distance function $d(R) = d(R; \xi^\dagger)$. Then, for arbitrary chosen $R_0 > 0$ the estimate*

$$D_{\xi^\dagger}(x_\alpha^\delta, x^\dagger) \leq K_1 \frac{\delta^p}{\alpha} + K_2 \alpha^{\frac{\kappa}{p-\kappa}} R^{\frac{p}{(1-c_2)(p-\kappa)}} + K_3 d(R) \tag{3.18}$$

holds for all $R \geq R_0$. Here, K_1, K_2 and K_3 are the constants of Lemma 3.2.2.

So, for fixed parameters α and δ we derived a family of error bounds. For proving convergence rates in the next section we choose one particular parameter $R = R(\alpha)$ in dependence of α in such a way that the last two terms of the estimate (3.18) are of the same order with respect to the regularization parameter α. By the monotonicity of the distance function this can be done in a unique manner as long as $d(R; \xi^\dagger) \to 0$ as $R \to \infty$.

3.2.3 Convergence rates

We present our first convergence rate result by assuming that the source condition (3.11) is fulfilled, see [40, Theorem 3.3]. For $\kappa = 1$ this theorem coincides with the convergence rates result in [45] and for $c_1 = 1$ with the analysis in [82].

Theorem 3.2.1 (Convergence rates I) *Assume (A1)-(A6) and $\xi^\dagger$ satisfies (3.11) for some $\omega \in \mathcal{Y}^*$. Moreover, let either $c_1 > 0$ or $L\|\omega\| < 1$ holds true. Then, an a-priori parameter choice $\alpha := \delta^{p-\kappa}$ leads to the convergence rate*

$$D_{\xi^\dagger}(x_\alpha^\delta, x^\dagger) = \mathcal{O}(\delta^\kappa) \quad \textit{as} \quad \delta \to 0. \tag{3.19}$$

PROOF. By assumption we have $\beta_1 < 1$. Hence, estimate (3.16) holds with $d = 0$. Now we choose $\alpha = \alpha(\delta)$ such that

$$\frac{\delta^p}{\alpha} = \alpha^{\frac{\kappa}{p-\kappa}} = \alpha^{\frac{p}{p-\kappa}-1} \Leftrightarrow \delta = \alpha^{\frac{1}{p-\kappa}} \Leftrightarrow \alpha = \delta^{p-\kappa}$$

which leads to the convergence rate (3.19). ■

For violated source condition (3.11) we have to exclude the case $c_1 = 0, c_2 = 1$ since the smallness condition $\beta_1 = L\|\omega\| < 1$ of Lemma 3.2.2 in this situation contradicts the idea of distance functions. A simple calculation shows that for $\delta \to 0$ (and hence $\alpha = \alpha(\delta) \to 0$ we suggest the choice $R \to \infty$. Hence the smallness condition will be violated for δ sufficiently small. For $c_2 \neq 1$ the following convergence rates result holds.

Theorem 3.2.2 (Convergence rates II) *Assume(A1)-(A6) and $c_1 > 0$. Suppose that $\xi^\dagger \in \overline{\mathcal{R}(F'(x^\dagger)^\star)} \setminus \mathcal{R}(F'(x^\dagger)^\star)$ has distance function $d(R) := d(R;\xi^\dagger)$. We define the functions*

$$\tilde{\Theta}(R) := d(R)^{\frac{p-\kappa}{\kappa}} R^{-\frac{p}{c_1}},\ \tilde{\Psi}(\alpha) := \left(\alpha\, d\left(\tilde{\Theta}^{-1}(\alpha)\right)\right)^{\frac{1}{p}} \quad \text{and} \quad \tilde{\Phi}(R) := d(R)^{\frac{1}{\kappa}} R^{-\frac{1}{c_1}}.$$

Then, for an a-priori choice $\alpha := \tilde{\Psi}^{-1}(\delta)$, we achieve the convergence rate

$$D_{\xi^\dagger}(x_\alpha^\delta, x^\dagger) = \mathcal{O}\left(d\left(\tilde{\Phi}^{-1}(\delta)\right)\right) \quad \text{as} \quad \delta \to 0. \tag{3.20}$$

PROOF. We apply the error estimates (3.18) of Corollary 3.2.1. By balancing the three terms of the estimates we get

$$\alpha^{\frac{\kappa}{p-\kappa}} = \frac{d(R)}{R^{\frac{p}{(1-c_2)(p-\kappa)}}} \Leftrightarrow \alpha = \frac{d(R)^{\frac{p-\kappa}{\kappa}}}{R^{\frac{p}{\kappa(1-c_2)}}} = \frac{d(R)^{\frac{p-\kappa}{\kappa}}}{R^{\frac{p}{c_1}}} = \tilde{\Theta}(R) \Leftrightarrow R = \tilde{\Theta}^{-1}(\alpha).$$

Moreover, we set

$$\frac{\delta^p}{\alpha} = d\left(\tilde{\Theta}^{-1}(\alpha)\right) \Leftrightarrow \delta = \left(\alpha\, d\left(\tilde{\Theta}^{-1}(\alpha)\right)\right)^{\frac{1}{p}} = \tilde{\Psi}(\alpha),$$

which describes the parameter choice $\alpha := \tilde{\Psi}^{-1}(\delta)$. Finally,

$$\delta = \alpha^{\frac{1}{p}} d(R)^{\frac{1}{p}} = \frac{d(R)^{\frac{p-\kappa}{p\kappa}+\frac{1}{p}}}{R^{\frac{1}{c_1}}} = d(R)^{\frac{1}{\kappa}} R^{-\frac{1}{c_1}} = \tilde{\Phi}(R)$$

holds, which provides $R = \tilde{\Phi}^{-1}(\delta)$. Hence, choosing $\alpha := \tilde{\Psi}^{-1}(\delta)$ we obtain the convergence rate (3.20). ■

Example 3.2.2 *We consider the special situation $c_1 = 1$ and $c_2 = 0$. Then $\kappa = 1$ and the functions $\tilde{\Theta}(R)$ and $\tilde{\Phi}(R)$ simplifies to*

$$\tilde{\Theta}(R) = d(R)^{p-1} R^{-p} \quad \text{and} \quad \tilde{\Phi}(R) = d(R)\, R^{-1}.$$

This was proven in [32, Theorem 7.2]. Moreover, even though the optimal parameter choice $\alpha := \tilde{\Psi}^{-1}(\delta)$ depends on the parameter p the convergence rates (3.19) and (3.20) are not affected by the specific choice of p. If $\mathcal{Y}$ is supposed to be a Hilbert space then we have $p = 2$ and $\tilde{\Theta}(R) = d(R)R^{-2}$.

We also consider the example of power-type distance functions.

Example 3.2.3 *We assume* $d(R) = \mathcal{C}\, R^{\frac{\mu}{\mu-1}}$ *for some* $0 < \mu < 1$*. The motivation for this specific structure was given in Section 2.4. Let* $\mathcal{C}$ *be a generic constant. Then we obtain*

$$\tilde{\Phi}(R) = d(R)^{\frac{1}{\kappa}} R^{-\frac{1}{c_1}} = \mathcal{C}\, R^{\frac{\mu(1-c_2)}{(\mu-1)c_1} - \frac{1}{c_1}} = \mathcal{C}\, R^{\frac{1-\mu c_2}{(\mu-1)c_2}}.$$

Hence, we can conclude

$$d\left(\tilde{\Phi}^{-1}(\delta)\right) = \mathcal{C}\, \delta^{\frac{\mu}{\mu-1}\frac{(\mu-1)c_1}{1-\mu c_2}} = \mathcal{C}\, \delta^{\frac{\mu c_1}{1-\mu c_2}}.$$

Sow we derive a convergence rate $D_{\xi^\dagger}(x_\alpha^\delta, x^\dagger) \sim \delta^{\frac{\mu c_1}{1-\mu c_2}}$ *as* $\delta \to 0$*. Since* $c_2 < 1$ *we have* $\frac{\mu c_1}{1-\mu c_2} \to 0$ *as* $\mu \to 0$ *and* $\frac{\mu c_1}{1-\mu c_2} \to \kappa$ *as* $\mu \to 1$*.*

Additional we can compare this rate with classical convergence rates for regularization approaches in Hilbert spaces.

Example 3.2.4 *We assume* $\mathcal{X}$ *and* $\mathcal{Y}$ *to be Hilbert spaces and* $P(x) = \frac{1}{2}\|x\|^2$*. Then we further derive*

$$\|x_\alpha^\delta - x^\dagger\| = D_{\xi^\dagger}(x_\alpha^\delta, x^\dagger)^{\frac{1}{2}} = \mathcal{O}\left(\delta^{\frac{\mu c_1}{2(1-\mu c_2)}}\right) \quad as \quad \delta \to 0.$$

In particlar, for linear operators we have $c_2 = 0$ *and* $c_1 = 1$*. So we obtain the rate* $\|x_\alpha^\delta - x^\dagger\| \sim \delta^{\frac{\mu}{2}}$ *as* $\delta \to 0$*. On the other hand, the power-type distance function implies a source condition* $x^\dagger \in \mathcal{R}((A^*A)^{\frac{\mu}{2}})$*. Then the optimal rate under this source condition is* $\|x_\alpha^\delta - x^\dagger\| \sim \delta^{\frac{\mu}{\mu+1}}$ *as* $\delta \to 0$ *which is better than the achieved one with the above calculations. This follows from the observation that in general we have no information about the relation of the behaviour of the norm* $\|x_\alpha^\delta - x^\dagger\|$ *and the Bregman distance* $D_{\xi^\dagger}(x_\alpha^\delta, x^\dagger)$ *for* $x_\alpha^\delta \to x^\dagger$*.*

3.2.4 On norm convergence

In order to prove convergence rates with respect to the norm we have to suppose an additional condition to the penalty functional P. In particular we deal with the following assumption, see e.g. [40].

(A7) The Bregman distance $D_{\xi^\dagger}(\cdot, x^\dagger)$ is s-coercive for some $2 \le s < \infty$, i.e. there exists a constant $C_s > 0$ such that

$$D_{\xi^\dagger}(x, x^\dagger) = P(x) - P(x^\dagger) - \langle \xi^\dagger, x - x^\dagger \rangle \ge C_s \|x - x^\dagger\|^s$$

for all $x \in \mathcal{S}_{\alpha_{max}, y}(\varrho_{max})$.

There is another consequence of this additional assumption. It turns out, that we can improve the convergence rates of the previous section. The reason can be found in the last term of the error bound (3.17). In the proof of this estimate we have replaced the term $\|x_\alpha^\delta - x^\dagger\|$ by a fixed bound ϱ_x from above. However, if we can ensure convergence $x_\alpha^\delta \to x^\dagger$ for $\delta \to 0$ then the term $d(R)\|x_\alpha^\delta - x^\dagger\|$ decays faster to zero than $d(R)$ which leads to higher convergence rates. That is exactly the point, which we want to prove in this section.

In a first step we can present the following error bound result.

Lemma 3.2.3 *Assume (A1)-(A7), $c_1 > 0$ and $\xi^\dagger \in \mathcal{M}_L(R, d)$ for some $d \geq 0$ and some $R \geq R_0 > 0$. Then there exist constants $\tilde{K}_1, \tilde{K}_2, \tilde{K}_3 > 0$ and $\hat{K}_1, \hat{K}_2, \hat{K}_3 > 0$ such that the estimates*

$$D_{\xi^\dagger}(x_\alpha^\delta, x^\dagger) \leq \tilde{K}_1 \frac{\delta^p}{\alpha} + \tilde{K}_2 \alpha^{\frac{\kappa}{p-\kappa}} R^{\frac{p}{(1-c_2)(p-\kappa)}} + \tilde{K}_3 d^{\frac{s}{s-1}} \tag{3.21}$$

and

$$\|x_\alpha^\delta - x^\dagger\| \leq \hat{K}_1 \frac{\delta^{\frac{p}{s}}}{\alpha^{\frac{1}{s}}} + \hat{K}_2 \alpha^{\frac{\kappa}{s(p-\kappa)}} R^{\frac{p}{s(1-c_2)(p-\kappa)}} + \hat{K}_3 d^{\frac{1}{s-1}} \tag{3.22}$$

hold. None of those constants depend on δ, α, R or d.

PROOF. From estimate (3.17) and (A7) we conclude

$$\begin{aligned}
D_{\xi^\dagger}(x_\alpha^\delta, x^\dagger) &\leq K_1 \frac{\delta^p}{\alpha} + K_2 \alpha^{\frac{\kappa}{p-\kappa}} R^{\frac{p}{(1-c_2)(p-\kappa)}} + K_3 \|x_\alpha^\delta - x^\dagger\| d \\
&\leq K_1 \frac{\delta^p}{\alpha} + K_2 \alpha^{\frac{\kappa}{p-\kappa}} R^{\frac{p}{(1-c_2)(p-\kappa)}} + K_3 \left(\frac{D_{\xi^\dagger}(x_\alpha^\delta, x^\dagger)}{C_s} \right)^{\frac{1}{s}} d \\
&\leq K_1 \frac{\delta^p}{\alpha} + K_2 \alpha^{\frac{\kappa}{p-\kappa}} R^{\frac{p}{(1-c_2)(p-\kappa)}} + \frac{1}{s} D_{\xi^\dagger}(x_\alpha^\delta, x^\dagger) \\
&\quad + \frac{s-1}{s} C_s^{\frac{1}{s-r}} K_3^{\frac{r}{s-1}} d^{\frac{s}{s-1}}
\end{aligned}$$

after applying (3.13) with $p_1 = q$, $p_2 = \frac{s}{s-1}$, $a = D_{\xi^\dagger}(x_\alpha^\delta, x^\dagger)$ and $b = C_s^{-\frac{1}{s}} K_3 d$. Hence we derive (3.21) with $\tilde{K}_1 = \frac{s}{s-1} K_1$, $\tilde{K}_2 = \frac{s}{s-1} K_2$ and $\tilde{K}_3 = C_s^{1/(1-s)} K_3^{s/(s-1)}$. The norm estimate (3.22) follows immediately from (3.21) and (A7) with $\hat{K}_i = (\tilde{K}_i / C_s)^{1/s}$, $i = 1, 2$, and $\hat{K}_3 = (\tilde{K}_3 / C_s)^{1/s} = (K_3 / C_s)^{1/(s-1)}$. ■

The following example illustrates the error bounds result and their dependency on the underlying parameters.

Example 3.2.5 *We consider again some special situations for (3.22).*

(a) If $c_1 = 1$, $c_2 = 0$ and (and hence $\kappa = 1$) we have

$$\|x_\alpha^\delta - x^\dagger\| \leq \hat{K}_1 \frac{\delta^{\frac{p}{s}}}{\alpha^{\frac{1}{s}}} + \hat{K}_2 \alpha^{\frac{1}{s(p-1)}} R^{\frac{p}{s(p-1)}} + \hat{K}_3 d^{\frac{1}{s-1}}.$$

(b) *If we further assume* $P(x) := \frac{1}{2}\|x - x^*\|^2$ *with some a priori guess* $x^* \in \mathcal{X}$ *and* $\mathcal{X}$ *is supposed to be a Hilbert space then* $s = 2$ *holds and*

$$\|x_\alpha^\delta - x^\dagger\| \le \hat{K}_1 \frac{\delta^{\frac{p}{2}}}{\alpha^{\frac{1}{2}}} + \hat{K}_2 \alpha^{\frac{1}{2(p-1)}} R^{\frac{p}{2(p-1)}} + \hat{K}_3 d.$$

Moreover, if we have chosen $p = 2$ *then this estimates further simplifies to*

$$\|x_\alpha^\delta - x^\dagger\| \le \hat{K}_1 \frac{\delta}{\alpha^{\frac{1}{2}}} + \hat{K}_2 \alpha^{\frac{1}{2}} R + \hat{K}_3 d$$

This, in particular, is the known structure of the error bound for Tikhonov regularization in Hilbert spaces, see e.g. [42] in the case of linear operators.

For the sake of completeness we prove an alternative error bound result, see [36].

Lemma 3.2.4 *Assume (A1)-(A7),* $c_1 > 0$ *and* $\xi^\dagger \in \mathcal{M}_L(R, d)$ *for some* $d \ge 0$ *and some* $R \ge R_0 > 0$. *Then the estimates*

$$D_{\xi^\dagger}(x_\alpha^\delta, x^\dagger) \le \max\left\{ 2\left(K_1 \frac{\delta^p}{\alpha} + K_2 \alpha^{\frac{\kappa}{p-\kappa}} R^{\frac{p}{(1-c_2)(p-\kappa)}}\right), (2K_3)^{\frac{s}{s-1}} \left(\frac{1}{C_s}\right)^{\frac{1}{2s-2}} d^{\frac{s}{s-1}} \right\}.$$

and

$$\|x_\alpha^\delta - x^\dagger\| \le \max\left\{ \left[\frac{2}{C_s}\left(K_1 \frac{\delta^p}{\alpha} + K_2 \alpha^{\frac{\kappa}{p-\kappa}} R^{\frac{p}{(1-c_2)(p-\kappa)}}\right)\right]^{\frac{1}{s}}, \left(\frac{2\,K_3}{C_s}\, d\right)^{\frac{1}{s-1}} \right\}.$$

hold. Here K_1, K_2 *and* K_3 *are the constants of Lemma 3.2.2.*

Proof. From the proof of Lemma 3.2.3 we derive

$$D_{\xi^\dagger}(x_\alpha^\delta, x^\dagger) \le K_1 \frac{\delta^p}{\alpha} + K_2 \alpha^{\frac{\kappa}{p-\kappa}} R^{\frac{p}{(1-c_2)(p-\kappa)}} + K_3 \left(\frac{1}{C_s}\right)^{\frac{1}{s}} d D_{\xi^\dagger}(x_\alpha^\delta, x^\dagger)^{\frac{1}{s}}.$$

Instead of Young's inequality we now apply the following implication, see e.g. [90, p.108]:

$$a,\, b,\, c > 0,\ 0 < q < 2,\ a^2 \le b^2 + c\, a^q \implies a \le \max\left\{\sqrt{2} b, (2c)^{\frac{1}{2-q}}\right\}.$$

We set

$$a := \sqrt{D_{\xi^\dagger}(x_\alpha^\delta, x^\dagger)}, \quad b := \sqrt{K_1 \frac{\delta^p}{\alpha} + K_2 \alpha^{\frac{\kappa}{p-\kappa}} R^{\frac{p}{(1-c_2)(p-\kappa)}}}, \quad c := K_3 \left(\frac{1}{C_s}\right)^{\frac{1}{s}} d$$

and $q := \frac{2}{s}$. Hence we derive

$$\sqrt{D_{\xi^\dagger}(x_\alpha^\delta, x^\dagger)} \le \max\left\{ \sqrt{2\left(K_1 \frac{\delta^p}{\alpha} + K_2 \alpha^{\frac{\kappa}{p-\kappa}} R^{\frac{p}{(1-c_2)(p-\kappa)}}\right)}, (2K_3)^{\frac{s}{2s-2}} \left(\frac{1}{C_s}\right)^{\frac{1}{2s-2}} d^{\frac{s}{2s-2}} \right\}.$$

Since $a \le \max\{b, c\}$ implies $a^2 \le \max\{b^2, c^2\}$ for $a, b, c \ge 0$ we proved the first result. For the second estimate we start with

$$\|x_\alpha^\delta - x^\dagger\|^s \le \frac{1}{C_s}\left(K_1 \frac{\delta^p}{\alpha} + K_2 \alpha^{\frac{\kappa}{p-\kappa}} R^{\frac{p}{(1-c_2)(p-\kappa)}}\right) + \frac{K_3}{C_s}\, d\, \|x_\alpha^\delta - x^\dagger\|.$$

Hence we set

$$a := \|x_\alpha^\delta - x^\dagger\|^{\frac{s}{2}}, \quad b := \left[\frac{1}{C_s}\left(K_1 \frac{\delta^p}{\alpha} + K_2 \alpha^{\frac{\kappa}{p-\kappa}} R^{\frac{p}{(1-c_2)(p-\kappa)}}\right)\right]^{\frac{1}{2}}, \quad c := \frac{K_3}{C_s} d$$

and $q := \frac{2}{s}$ which gives

$$\|x_\alpha^\delta - x^\dagger\|^{\frac{s}{2}} \le \max\left\{\left[\frac{2}{C_s}\left(K_1 \frac{\delta^p}{\alpha} + K_2 \alpha^{\frac{\kappa}{p-\kappa}} R^{\frac{p}{(1-c_2)(p-\kappa)}}\right)\right]^{\frac{1}{2}}, \left(\frac{2\,K_3}{C_s} d\right)^{\frac{s}{2s-2}}\right\}.$$

This proves the second estimate. ■

We emphasize that the estimates of Lemma 3.2.3 and 3.2.4 are basically of the same order with respect to δ, α, R and d. We derive now an improved convergence rates result, see [40, Theorem 4.6 and Corollary 4.7].

Theorem 3.2.3 (Convergence rates III) *Assume (A1)-(A7) and $c_1 > 0$. Suppose that $\xi^\dagger \in \overline{\mathcal{R}(F'(x^\dagger)^\star)} \setminus \mathcal{R}(F'(x^\dagger)^\star)$ has distance function $d(R) := d(R; \xi^\dagger)$. We define the functions*

$$\Theta(R) := \frac{d(R)^{\frac{s(p-\kappa)}{(s-1)\kappa}}}{R^{\frac{p}{c_1}}}, \; \Psi(\alpha) := \alpha^{\frac{1}{p}} d\left(\Theta^{-1}(\alpha)\right)^{\frac{s}{(s-1)p}} \quad \text{and} \quad \Phi(R) := d(R)^{\frac{s}{(s-1)\kappa}} R^{-\frac{1}{c_1}}.$$

Then, for an a-priori choice $\alpha := \Psi^{-1}(\delta)$, we achieve the convergence rates

$$D_{\xi^\dagger}(x_\alpha^\delta, x^\dagger) = \mathcal{O}\left(d\left(\Phi^{-1}(\delta)\right)^{\frac{s}{s-1}}\right) \tag{3.23}$$

and

$$\|x_\alpha^\delta - x^\dagger\| = \mathcal{O}\left(d\left(\Phi^{-1}(\delta)\right)^{\frac{1}{s-1}}\right) \quad \text{as} \quad \delta \to 0. \tag{3.24}$$

PROOF. We apply the same balancing idea as in Theorem 3.2.2. Here, the choice of R is described by

$$\alpha^{\frac{\kappa}{p-\kappa}} = \frac{d(R)^{\frac{s}{s-1}}}{R^{\frac{p}{(1-c_2)(p-\kappa)}}} \Leftrightarrow \alpha = \frac{d(R)^{\frac{r(p-\kappa)}{(s-1)\kappa}}}{R^{\frac{p}{\kappa(1-c_2)}}} = \frac{d(R)^{\frac{s(p-\kappa)}{(s-1)\kappa}}}{R^{\frac{p}{c_1}}} = \Theta(R) \Leftrightarrow R = \Theta^{-1}(\alpha).$$

The parameter choice $\alpha = \Psi^{-1}(\delta)$ we get from

$$\frac{\delta^p}{\alpha} = d\left(\Theta^{-1}(\alpha)\right)^{\frac{s}{s-1}} \Leftrightarrow \delta = \alpha^{\frac{1}{p}} d\left(\Theta^{-1}(\alpha)\right)^{\frac{s}{(s-1)p}} = \Psi(\alpha).$$

For the convergence rate we derive

$$\delta = \alpha^{\frac{1}{p}} d(R)^{\frac{s}{(s-1)p}} = d(R)^{\frac{s}{(s-1)p}+\frac{r(p-\kappa)}{(s-1)p\kappa}} R^{-\frac{1}{c_1}} = d(R)^{\frac{s}{(s-1)\kappa}} R^{-\frac{1}{c_1}} = \Phi(R).$$

This proves the rates (3.23) and (3.24). ■

The following example illustrates these results for some particular parameters combinations.

Example 3.2.6 *Again we consider some specific situations.*

(a) We assume $c_1 = 1$ and $c_2 = 0$. Then we have $\kappa = 1$ and $\Theta(R) = d(R)^{\frac{s(p-1)}{s-1}} R^{-p}$ and $\Phi(R) = d(R)^{\frac{s}{s-1}} R^{-1}$.

(b) If, additionally, $s = 2$, then these functions simplifies to $\Theta(R) = d(R)^{2(p-1)} R^{-p}$ and $\Phi(R) = d(R)^2 R^{-1}$. This was the situation considered in [32, Theorem 5.2].

(c) On the other hand, if $p = 2$ then we have $\Theta(R) = d(R)^{\frac{s-1}{s}} R^{-2}$. The function $\Phi(R)$, $R \geq 0$, which describes the convergence rates does not depend on the specific choice of p.

(d) If $p = s = 2$ then we obtain $\Theta(R) = d(R)^2 R^{-2}$.

More specific, we apply the convergence rates result to power-type distance functions.

Example 3.2.7 *We again assume $d(R) = \mathcal{C} R^{\frac{\mu}{\mu-1}}$ for some $0 < \mu < 1$. Let $\mathcal{C}$ be an generic constant. Then we obtain*

$$\Phi(R) = d(R)^{\frac{s}{(s-1)\kappa}} R^{-\frac{1}{c_1}} = \mathcal{C} R^{\frac{s\mu(1-c_2)}{(s-1)(\mu-1)c_1} - \frac{1}{c_1}} = \mathcal{C} R^{\frac{s+\mu-1-s\mu c_2}{(s-1)(\mu-1)c_2}}.$$

Hence, we can conclude

$$d\left(\Phi^{-1}(\delta)\right)^{\frac{1}{s-1}} = \mathcal{C}\, \delta^{\frac{1}{s-1}\frac{\mu}{\mu-1}\frac{(s-1)(\mu-1)c_1}{s+\mu-1-s\mu c_2}} = \mathcal{C}\, \delta^{\frac{\mu c_1}{\mu+s-1-s\mu c_2}}.$$

Sow we derive a convergence rate $\|x_\alpha^\delta - x^\dagger\| \sim \delta^{\frac{\mu c_1}{\mu+s-1-\mu c_2}}$ as $\delta \to 0$. This rate is in fact better than the rate of Example 3.2.3. There, under the additional coercivity we have the convergence rate $\|x_\alpha^\delta - x^\dagger\| \sim \delta^{\frac{\mu c_1}{s(1-\mu c_2)}}$ with respect to the norm. The quotient of both exponents is

$$\frac{\mu c_1}{\mu + s - 1 - s\mu c_2} \frac{s(1-\mu c_2)}{\mu c_1} = \frac{s(1-\mu c_2)}{s(1-\mu c_2) - 1 + \mu} > 1$$

since $\mu < 1$ was supposed.

The final example compares these estimates with the classical theory in Hilbert spaces.

Example 3.2.8 *We again assume $\mathcal{X}$ and $\mathcal{Y}$ to be Hilbert spaces and $P(x) = \frac{1}{2}\|x\|^2$. Hence we have $s = 2$. Then we further derive*

$$\|x_\alpha^\delta - x^\dagger\| = \mathcal{O}\left(\delta^{\frac{\mu c_1}{\mu+1-2\mu c_2}}\right) \quad \text{as} \quad \delta \to 0.$$

In particlar, for linear operators with $c_2 = 0$ and $c_1 = 1$ we achieve the rate $\|x_\alpha^\delta - x^\dagger\| \sim \delta^{\frac{\mu}{\mu+1}}$ as $\delta \to 0$ which is the optimal rate under the source condition $x^\dagger \in \mathcal{R}((A^*A)^{\frac{\mu}{2}})$.

3.2.5 On exact penalization

We consider the case $p = 1$, which was excluded in the previous sections. It is also known as exact penalization. Convergence rates for this case were firstly proven in [12] under assumption of the validity of the source condition (3.11). In particular, the setting $p = \kappa = 1$ plays a singular role since we cannot apply Young's inequality (3.13) in the proof of Lemma 3.2.2. However, as we will see, we can obtain similar convergence rates results also in this situation. We refer also to [32] where the case $P(x) := \frac{1}{2}\|x\|^2$ was studied.

We again start with an error bound result. As we will see for $p = \kappa = 1$ we have to distinguish between $\alpha > \beta_2^{-1}$ and $\alpha \leq \beta_2^{-1}$. We only consider the latter case which is applied in the further convergence rate analysis.

Lemma 3.2.5 *Assume $p = 1$, (A1)-(A6), $\beta_1 < 1$ and $\xi^\dagger \in \mathcal{M}(R, d)$ for some $R, d \geq 0$. If $\kappa < 1$ then the estimate (3.16) holds. For $\kappa = p = 1$ and $\alpha \leq \beta_2^{-1}$ we have*

$$D_{\xi^\dagger}(x_\alpha^\delta, x^\dagger) \leq \frac{1}{1-\beta_1}\left(\frac{2\delta}{\alpha} + \varrho_x d\right). \tag{3.25}$$

PROOF. Studying the proof of Lemma 3.2.2 we observe that estimate (3.16) holds as long as $\frac{p}{\kappa} > 1$ holds. For $p = \kappa = 1$ we conclude from the proof of Lemma 3.2.2 that

$$\|F(x_\alpha^\delta) - y^\delta\| + (1-\beta_1)\alpha D_{\xi^\dagger}(x_\alpha^\delta, x^\dagger) \leq \delta + \alpha\beta_2\|F(x_\alpha^\delta) - y^\delta\| + \alpha\beta_2\delta + \alpha\|x_\alpha^\delta - x^\dagger\|d.$$

By assumption we have $\alpha\beta_2 \leq 1$. Then we can drop all terms containing $\|F(x_\alpha^\delta) - y^\delta\|$ which proves the second case. ■

In particular, for the case $\kappa < 1$ all error bound and convergence rates results of the previous sections can be extended to the choice $p = 1$. Only for $\kappa = 1$ we have to perform extra considerations. In that case we cannot apply Young's inequality as in the proof of Lemma 3.2.2 for proving the accordant error bound. We present the corresponding convergence rates results, which are now an immediate consequence of this observation.

Theorem 3.2.4 (Convergence rates IV) *Assume (A1)-(A6) and $\xi^\dagger$ satisfies the source condition (3.11) for some $\omega \in \mathcal{Y}^*$. Moreover, let either $c_1 > 0$ or $L\|\omega\| < 1$. Then an a-priori parameter choice $\alpha := \delta^{1-\kappa}$ for $\kappa < 1$ or $\alpha \leq \beta_2$ for $\kappa = 1$ leads to the convergence rate (3.19).*

The result for $\kappa = 1$ highlights the exceptional position of the case $p = 1$: as opposite to $p > 1$ we can here (under assumption of the source condition (3.11)) achieve the convergence rates by choosing the regularization parameter α sufficiently small but independently of the noise level $\delta > 0$. It also turns out, that the choice $\alpha := \frac{1}{\beta_2}$ yields the best error estimate.

For violated source condition we again have to assume $c_1 > 0$. If $\kappa < 1$ then we can directly apply the convergence rates results of Theorem 3.2.2 and 3.2.3. So we only have to discuss the case $\kappa = 1$.

Theorem 3.2.5 (Convergence rates V) *Assume (A1)-(A6), $p = \kappa = 1$, $c_1 > 0$ and $\xi^\dagger \in \overline{\mathcal{R}(F'(x^\dagger)^\star)} \setminus \mathcal{R}(F'(x^\dagger)^\star)$ has distance function $d(R) := d(R; \xi^\dagger)$. We set $\tilde{\Theta}(R) := (c_1 L + R_0^{-c_2/c_1})^{-1} R^{-1/c_1}$ for some $R_0 > 0$.*

(i) With the functions $\tilde{\Psi}(\alpha)$ and $\tilde{\Phi}(R)$ as in Theorem 3.2.2 and the a-priori choice $\alpha := \tilde{\Psi}^{-1}(\delta)$ we achieve the convergence rate (3.20).

(ii) Let additionally the coercivity assumption (A7) hold and the functions $\Psi(\alpha)$ and $\Phi(R)$ are chosen as in Theorem 3.2.3 and $\Theta(R) = \tilde{\Theta}(R)$ Then an a-priori choice $\alpha = \Psi^{-1}(\delta)$ yields the convergence rates (3.23) and (3.24) respectively.

PROOF. From Lemma 3.2.1 and the definition of β_2 we have with $\kappa = 1$

$$\beta_2 = (1 - c_2)(L\,R)^{\frac{1}{1-c_2}} + R \leq \left(c_1 L + R_0^{\frac{-c_2}{c_1}}\right) R^{\frac{1}{c_1}}.$$

for $R \geq R_0$. Here we used that $\kappa = 1$ implies $c_1 = 1 - c_2$. Now we can follow the balancing steps in the proof of Theorem 3.2.2. We have chosen $R = \tilde{\Theta}^{-1}(\alpha)$ such that $\alpha\beta_2 \leq 1$ holds as long $R \geq R_0$. So we can apply the estimate (3.25) to obtain

$$D_{\xi^\dagger}(x_\alpha^\delta, x^\dagger) \leq \frac{1}{1-\beta_1}\left(2\left(c_1 L + R_0^{-\frac{c_2}{c_1}}\right)\delta\, R^{\frac{1}{c_1}} + \varrho_x d(R)\right).$$

The choice $\delta = \tilde{\Phi}(R) = d(R) R^{\frac{-1}{c_1}}$ implies the convergence rate (3.20). We only have to note, that we always can guarantee $R \geq R_0$ for δ sufficiently small. The second part of the proof can be treated analogously. ■

Example 3.2.9 *For $c_1 = 1$, $c_2 = 0$ we have $\kappa = 1$ and $\beta_2 = (1 + L)R$. Then the function $\tilde{\Theta}(R)$ simplifies to $\tilde{\Theta}(R) = [(1 + L)R]^{-1}$ which was observed in [32, Theorem 3.1]. For linear operators we have $L = 0$ and hence $\tilde{\Theta}(R) = R^{-1}$, see also [32, Theorem 5.1].*

3.2.6 A-posteriori parameter choices

The convergence rates of the previous sections are based on the knowledge of the distance functions and a proper a-priori choice of the regularization parameter α. We now discuss strategies how to choose the regularization parameter α without knowing the corresponding distance function by keeping the optimal convergence rates. We only discuss the case $p > \kappa$ here, i.e. the situation $p = \kappa = 1$ is excluded here. However, we point out that similar considerations shows the applicability of these parameter choices also for the case $p = \kappa = 1$.

a) The Discrepancy principle

If we consider classical Tikhonov regularization in Hilbert spaces $\mathcal{X}$ and $\mathcal{Y}$ for linear operator equations $Ax = y$, i.e. we have $p = 2$ and $P(x) = \frac{1}{2}\|x\|^2$, then the well-known

(Morozov-) discrepancy principle [71] suggests to choose the regularization parameter $\alpha = \alpha_D$ such that

$$\|Ax^\delta_{\alpha_D} - y^\delta\| = \delta. \tag{3.26}$$

In particular, we can find for each $0 < \delta < \|y\|$ and each $y^\delta \in \mathcal{Y}$ with $\|y - y^\delta\| \leq \delta$ a parameter α_D such that equality (3.26) is satisfied, see e.g. [21, Chapter 5.1]. However, for nonlinear equations this will not hold true even in Hilbert spaces, see [21, Chapter 10.3] and [59]. Even if such parameter $\alpha = \alpha_D$ exists, it is difficult to calculate them by a numerical procedure. Therefore different modifications of the discrepancy principle has been established for practical implementations. We suggest to apply the following one.

Definition 3.2.2 *Let be $\tau_0 > 1$. We choose $\alpha = \alpha_D > 0$ such that the estimates*

$$\delta \leq \|F(x^\delta_{\alpha_D}) - y^\delta\| \leq \tau_0\, \delta \tag{3.27}$$

hold. Then $x^\delta_\alpha := x^\delta_{\alpha_D}$ is chosen as regularized solution of (3.2).

Of course, the existence of such parameter α_D is not guaranteed for all situations. In particular if τ_0 is chosen too small the existence of α_D might be violated. However, in most practical application this version of the discrepancy principle has been established as successful tool for choosing the regularization parameter α in a proper way. To overcome the gap between theoretical predications and practical observations we here assume the existence of such regularization parameter α_D satisfying (3.27) explicitly. We again refer to [21, Chapter 10.3] for some modifications of the regularization approach (3.4) yielding the existence of such parameter (in Hilbert spaces). We present now the following convergence rates result.

Theorem 3.2.6 *Assume (A1)-(A6) and there exists a parameter α_D such that (3.27) holds.*

(i) If $\xi^\dagger$ satisfies source condition (3.11) for some $\omega \in \mathcal{Y}^*$, then the convergence rate (3.19) holds, whenever $c_1 > 0$ or $L\|\omega\| < 1$.

(ii) If $c_1 > 0$ and $\xi^\dagger \in \overline{\mathcal{R}(F'(x^\dagger)^\star)} \setminus \mathcal{R}(F'(x^\dagger)^\star)$ has distance function $d(R) = d(R;\xi^\dagger)$, then the rate (3.20) holds. Moreover, if additional (A7) holds true, then we achieve the convergence rate (3.23) and (3.24) respectively.

PROOF. From the proof of Lemma 3.2.2 we derive

$$\frac{1}{p}\|F(x^\delta_\alpha) - y^\delta\|^p + (1-\beta_1)\alpha D_{\xi^\dagger}(x^\delta_\alpha, x^\dagger) \leq \frac{\delta^p}{p} + \alpha\beta_2\|F(x^\delta_\alpha) - y^\delta\|^\kappa + \alpha\beta_2\delta^\kappa + \alpha\varrho_x d.$$

We subtract $\frac{\delta^p}{p}$ and use the estimate $\delta \leq \|F(x^\delta_\alpha) - y^\delta\| \leq \tau_0\delta$ to obtain

$$D_{\xi^\dagger}(x^\delta_\alpha, x^\dagger) \leq \frac{1}{1-\beta_1}\left(\beta_2(1+\tau_0)\delta^\kappa + \|x^\delta_\alpha - x^\dagger\| d\right).$$

If $\xi^\dagger \in \mathcal{R}(F'(x^\dagger)^\star)$ then we can set $d = 0$ for β_2 sufficiently large. This proves the first part. For violated source condition (3.11) and $c_1 > 0$ we again have $\beta_2 \le C\, R^{\frac{1}{1-c_2}}$ for some constant $C > 0$ and $R \ge R_0$. Moreover, $\|x_\alpha^\delta - x^\dagger\| \le \varrho_x$ holds. Hence, choosing R such that

$$R^{\frac{1}{1-c_2}}\delta^\kappa = d(R) \;\Leftrightarrow\; \delta = d(R)^{\frac{1}{\kappa}} R^{-\frac{1}{c_1}} \;\Leftrightarrow\; R = \tilde{\Phi}^{-1}(\delta)$$

we obtain the convergence rate (3.20). Under the additional coercivity assumption (A7) we derive the rates (3.23) and (3.24) by following the idea of the proof of Lemma 3.2.3. ■

The theorem shows in particular the following: for achieving the convergence rates of Section 3.2.3 and 3.2.4 we had to distinguish between the situations of satisfied and violated source condition (3.11) since different a-priori parameter choice strategies were applied. This is not necessary anymore for the a-posteriori parameter choice based on the discrepancy principle. This observation is important for the practical application of this method: the implementation of the discrepancy principle can be carried out without any knowledge of the (possible) validity of the source condition (3.11).

b) The Balancing principle

We know discuss the applicability of the balancing principle. The main theoretical background of this method we shortly repeated in the Appendix A.

Since Bregman distances are in general neither transitive nor symmetrically, they do not define a metric generally. So, the balancing principle cannot be applied to the error bounds in terms of Bregman distances. Therefore, as opposite to the discrepancy principle, the coercivity assumptions (A7) is necessary for the adaptability of this method.

Let the assumption (A7) be fulfilled and $c_1 > 0$. For satisfied source condition (3.11) and estimate (3.17) with $d = 0$ we derive

$$\|x_\alpha^\delta - x^\dagger\| \le \left(\frac{K_1}{C_s}\right)^{\frac{1}{s}} \frac{\delta^{\frac{p}{s}}}{\alpha^{\frac{1}{s}}} + K_2 \alpha^{\frac{\kappa}{s(p-\kappa)}} R^{\frac{p}{s(1-c_2)(p-\kappa)}}.$$

On the other hand, for $\xi^\dagger \in \overline{\mathcal{R}(F'(x^\dagger)^\star)} \setminus \mathcal{R}(F'(x^\dagger)^\star)$ with distance function $d(R) = d(R;\xi^\dagger)$ we derive from the estimate (3.22)

$$\begin{aligned}
\|x_\alpha^\delta - x^\dagger\| &\le \hat{K}_1 \frac{\delta^{\frac{p}{s}}}{\alpha^{\frac{1}{s}}} + \hat{K}_2 \alpha^{\frac{\kappa}{s(p-\kappa)}} R^{\frac{p}{r(1-c_2)(p-\kappa)}} + \hat{K}_3 d(R)^{\frac{1}{s-1}} \\
&= \left[\frac{s}{s-1}\frac{K_1}{C_s}\right]^{\frac{1}{s}} \frac{\delta^{\frac{p}{s}}}{\alpha^{\frac{1}{s}}} + \left(\hat{K}_2 + \hat{K}_3\right) d\left(\Theta^{-1}(\alpha)\right)^{\frac{1}{s-1}}
\end{aligned}$$

after the first balancing step of Theorem 3.2.3. Hence, in both cases we have the same error structure. The first part of these error estimates is (up to a constant) independent of the unknown (approximate) source conditions whereas the latter part do not depend on the noise level δ. So we can apply the balancing principle with

$$\psi(\alpha) := \left[\frac{s}{s-1}\frac{K_1}{C_s}\right]^{\frac{1}{s}} \frac{\delta^{\frac{p}{s}}}{\alpha^{\frac{1}{s}}}.$$

Moreover, we observe, that for $q > 0$ and all $\alpha > 0$

$$\psi(\alpha) = q^{\frac{1}{s}} \psi(q\,\alpha)$$

holds. In particular, the additional assumption of Proposition A.0.1 holds with $D = q^{\frac{1}{s}}$.

3.3 High-order convergence rates

3.3.1 Motivation

In the previous section the reference source condition (3.11) plays the role of a limit situation: with the subsequent technique we cannot prove faster convergence rates than $D_{\xi^\dagger}(x_\alpha^\delta, x^\dagger) = \mathcal{O}(\delta^\kappa)$ and $\|x_\alpha^\delta - x^\dagger\| = \mathcal{O}(\delta^{\frac{\kappa}{s}})$ as $\delta \to 0$ which was the optimal rates when the source condition (3.11) was fulfilled. We motivate the following section with the underlying example.

Example 3.3.1 *We can apply the convergence rates results to classical Tikhonov regularization for linear equations in Hilbert spaces $\mathcal{X}$ and $\mathcal{Y}$, i.e. we have $\kappa = 1$, $P(x) = \frac{1}{2}\|x\|^2$ and $p = s = 2$. Then reference source condition (3.11) says $x^\dagger \in \mathcal{R}(A^*)$ and a parameter choice $\alpha \sim \delta$ implies the convergence rate $\|x_\alpha^\delta - x^\dagger\| = \mathcal{O}(\sqrt{\delta})$ as $\delta \to 0$. However, it is well-known that we can achieve higher convergence rates by assuming a stronger source condition. In particular, for $x^\dagger \in \mathcal{R}(A^*A)$ and parameter choice $\alpha \sim \delta^{\frac{2}{3}}$ we can prove a convergence rate $\|x_\alpha^\delta - x^\dagger\| = \mathcal{O}(\delta^{\frac{2}{3}})$ as $\delta \to 0$, see e.g. [21, Section 5.1]. The same rate was proved for nonlinear equations and penalty term $P(x) := \frac{1}{2}\|x - x_*\|^2$ by assuming F to be nonlinear of degree $(0,1)$ and applying the source condition $x^\dagger - x_* \in \mathcal{R}(F'(x^\dagger)^* F'(x^\dagger))$, see e.g. [21, Theorem 10.7]. Here, $x_* \in \mathcal{X}$ is an arbitrary chosen inital guess.*

Therefore we deal in this section with the following questions:

1. Can we find a (reference) source condition which transfers the Hilbert space condition $x^\dagger \in \mathcal{R}(A^*A)$ respectively $x^\dagger - x_* \in \mathcal{R}(F'(x^\dagger)^* F'(x^\dagger))$ for classical Tikhonov regularization to the general regularization setting in Banach spaces?

2. Under what kind of conditions on the penalty functional P (and the spaces $\mathcal{X}$ and $\mathcal{Y}$ under considerations) we can achieve the rate $\|x_\alpha^\delta - x^\dagger\| = \mathcal{O}(\delta^{\frac{2}{3}})$ as $\delta \to 0$ and how does the corresponding parameter choice $\alpha = \alpha(\delta)$ looks like?

3. If the new reference source condition is violated - can we adopt the concept of distance functions of the previous section as measure of the violation of the reference source condition in order to prove convergence rates?

The last point seems to be the most remarkable one. In particular, an one-to-one transfer of the convergence rates theory of Section 3.2 on the basis of a stronger reference source condition than (3.11) and corresponding distance functions would render the previous section unnecessary since all presented convergence rates results were included in this new situation. However, it turns out, that the below analysis needs stronger assumptions for proving error bound and convergence rates, namely:

1. We can apply the idea of distance functions for proving convergence rates only for linear equations, i.e. nonlinear equations are excluded in this part of the discussion.

2. Even in the linear case we need some additional assumptions. In particular, the smoothness of the penalty functional P as well as the smoothness of the space $\mathcal{Y}$ have influence on the error bounds and underlying convergence rates results.

Since the additional conditions are essential and restrictive this oberservation gives the qualification for a separate consideration of the previous and the present section. Therefore we speak of the situation of *low order convergence rates* in the previous and of *high order convergence rates* in the present section. However, under the assumption of certain smoothness conditions the results of Section 3.2 are included in the higher order situation.

3.3.2 Some preliminary estimates

It turns out in the subsequent analysis that we have to restrict our considerations to nonlinear operators F with degree $(0,1)$ of nonlinearity, i.e. we have the estimate

$$\|F(x) - F(x^\dagger) - F'(x^\dagger)(x - x^\dagger)\| \leq L\, D_{\xi^\dagger}(x, x^\dagger), \qquad \forall\, x \in \mathcal{S}_{\alpha_{max},y}(\varrho_{max}), \tag{3.28}$$

which holds for some constant $L > 0$. To shorten the notation we set $G := F'(x^\dagger)$. As already mentioned we need additional smoothness assumptions on the penalty functional P and the space $\mathcal{Y}$. These conditions are stated below.

(A8) There exists a constant $G_r > 0$ and an exponent $1 < r \leq 2$ such that

$$D_{\xi^\dagger}(x, x^\dagger) \leq G_r \|x - x^\dagger\|^r$$

holds for all $x \in \mathcal{S}_{\alpha_{max},y}(\varrho_{max})$.

(A9) The space $\mathcal{X}$ is supposed to be reflexive and the space $\mathcal{Y}$ is a-smooth for some $1 < a \leq 2$.

In a second step we have to introduce a new reference source condition which generalizes the condition $x^\dagger - x_* \in \mathcal{R}(F'(x^\dagger)^* F'(x^\dagger))$ in Hilbert spaces for classical Tikhonov regularization of nonlinear equations. We recall the definition of the Tikhonov functional $T_{\alpha,y^\delta}(x) := \frac{1}{p}\|F(x) - y^\delta\|^p + \alpha\, P(x)$ under consideration. Let us first assume $p > 1$. Then the most intuitive choice of such source condition would be

$$\xi^\dagger = G^\star J_p(G\,\omega), \qquad \omega \in \mathcal{X},\ \|\omega\| \leq R^{\frac{1}{p-1}}, \tag{3.29}$$

for some $R \geq 0$.The motivation of restricting the norm of the source element $\omega \in \mathcal{X}$ by $R^{\frac{1}{p-1}}$ is given in the next section in connection with a proper definition of appropriate distance functions. Here, the exponent p of the residual term of the Tikhonov functional determines the choice of the specific duality mapping $J_p : \mathcal{Y} \longrightarrow \mathcal{Y}^*$. However, the validity of such source condition does not depend on the specific choice of the parameter p. In particular, $\xi^\dagger = G^\star J_p(G\,\omega)$ for some $\omega \in \mathcal{X}$ implies $\xi^\dagger = G^\star J_2(G\,\tilde{\omega})$ with $\tilde{\omega} := \|G\,\omega\|^{p-2}\omega$. Introducing the source condition (3.29) in the way, which is presented here, has technical reasons leading to a more coherent analysis. The operator $\omega \mapsto G^\star J_p(G\,\omega)$ on the right hand side is in general nonlinear which might be rather unusual in connection with source conditions at the first glance. However, we will show that exactly this condition (3.29) provides the desired convergence rates.

We start with an example to illustrate condition (3.29) for a specific situation.

Example 3.3.2 *We return to the linear Example 1.2.1. We set $P(x) := \frac{1}{q}\|x\|^q$, $q > 1$. We assume $\omega \equiv 1$ to be the underlying source element. Hence, for $t \in [0,1]$ we can calculate that*

$$x^\dagger(t) = (J_q)^{-1}\left(A^\star J_p(A\,\omega)\right)(t) = \left[\int_t^1 \left(\int_0^{\hat{t}} 1\,d\tau\right)^{p-1} d\hat{t}\right]^{\frac{1}{q-1}} = \left[\frac{1}{p}(1-t^p)\right]^{\frac{1}{q-1}}$$

is the corresponding solution of (3.3) which satisfies the source condition (3.29) with $R \geq 1$. The source condition contains smoothness assumption and boundary conditions.

As in the previous section we follow the strategy of approximate source conditions (and distance functions) again. Here, the source condition (3.29) plays now the role of the reference source condition which is allowed to be violated. Therefore we weaken the source condition (3.29) by assuming an approximative source condition

$$\xi^\dagger = G^\star J_p(G\,\omega) + \upsilon, \qquad \omega \in \mathcal{X},\ \upsilon \in \mathcal{X}^*,\ \|\omega\| \leq R^{\frac{1}{p-1}},\ \|\upsilon\| \leq d. \tag{3.30}$$

for some $R, d \geq 0$. For presenting a unified framework we introduce the sets

$$\mathcal{M}_H(R,d) := \left\{\xi \in \mathcal{X}^* \ :\ \xi = G^\star J_p(G\,\omega) + \upsilon,\ \omega \in \mathcal{X},\ \upsilon \in \mathcal{X}^*,\ \|\omega\| \leq R^{\frac{1}{p-1}},\ \|\upsilon\| \leq d\right\}.$$

for each $R, d \geq 0$. Again, the number $d \geq 0$ describes the maximal violation of the (reference) source condition (3.29) when the norm of the source element $\omega \in \mathcal{X}$ is bounded by $R^{\frac{1}{p-1}}$. Additionally we introduce the notation

$$D_{Y,p}(y_1, y_2) := \frac{1}{p}\|y_1\|^p - \frac{1}{p}\|y_2\|^p - \langle J_p(y_2), y_1 - y_2\rangle, \quad y_1, y_2 \in \mathcal{Y}.$$

Now we can prove a first general error estimate which does not depend on the additional smoothness conditions (A8) and (A9), see also [35, Lemma 6.1].

Lemma 3.3.1 *Suppose $p > 1$. Assume (A1)-(A6), (3.28) and $\xi^\dagger \in \partial P(x^\dagger)$ satisfies (3.30) for some $\omega \in \mathcal{X}$ and $\upsilon \in \mathcal{X}^*$. If $L\,\|G\,\omega\|^{p-1} < 1$, $\gamma := \alpha^{\frac{1}{p-1}}$ and $x^\dagger - \gamma\,\omega \in \mathcal{D}$ then* *the estimate*

$$(1 - L\,\|G\,\omega\|^{p-1})\,D_{\xi^\dagger}(x_\alpha^\delta, x^\dagger) \le \big(1 + L\,\|G\,\omega\|^{p-1}\big)\,D_{\xi^\dagger}(x^\dagger - \gamma\omega, x^\dagger) + \|v\|\,(\varrho_x + \gamma\|\omega\|) + \frac{1}{\alpha} D_{Y,p}(y^\delta - F(x^\dagger - \gamma\omega), \gamma\,F'(x^\dagger)\,\omega) \tag{3.31}$$

holds.

PROOF. For $\gamma > 0$ we have

$$\begin{aligned} P(x_\alpha^\delta) - P(x^\dagger - \gamma\omega) &= P(x_\alpha^\delta) - P(x^\dagger) + P(x^\dagger) - P(x^\dagger - \gamma\omega) \\ &= \langle \xi^\dagger, x_\alpha^\delta - x^\dagger\rangle + D_{\xi^\dagger}(x_\alpha^\delta, x^\dagger) - \langle \xi^\dagger, -\gamma\omega\rangle - D_{\xi^\dagger}(x^\dagger - \gamma\omega, x^\dagger) \\ &= \langle J_p(G\,\omega), G(x_\alpha^\delta + \gamma\omega - x^\dagger)\rangle + D_{\xi^\dagger}(x_\alpha^\delta, x^\dagger) - D_{\xi^\dagger}(x^\dagger - \gamma\omega, x^\dagger) \\ &\quad + \langle v, x_\alpha^\delta + \gamma\omega - x^\dagger\rangle \\ &= \langle J_p(G\,\omega), G(x_\alpha^\delta - x^\dagger)\rangle + D_{\xi^\dagger}(x_\alpha^\delta, x^\dagger) - D_{\xi^\dagger}(x^\dagger - \gamma\omega, x^\dagger) \\ &\quad + \gamma\|G\,\omega\|^p + \langle v, x_\alpha^\delta + \gamma\omega - x^\dagger\rangle. \end{aligned}$$

Here, we applied $\langle J_p(G\,\omega), G\,\omega\rangle = \|G\,\omega\|^p$. We introduce the notations

$$R_1 := F(x_\alpha^\delta) - F(x^\dagger) - G(x_\alpha^\delta - x^\dagger) \quad \text{and} \quad R_2 := F(x^\dagger - \gamma\,\omega) - F(x^\dagger) + \gamma\,G\,\omega.$$

By assumptions, the estimates $\|R_1\| \le L\,D_{\xi^\dagger}(x_\alpha^\delta, x^\dagger)$ and $\|R_2\| \le L\,D_{\xi^\dagger}(x^\dagger - \gamma\,\omega, x^\dagger)$ hold. Moreover, with $J_p(\gamma\,G\,\omega) = \gamma^{p-1} J_p(G\,\omega)$ we obtain

$$\begin{aligned} \frac{1}{p}\|F(x^\dagger - \gamma\,\omega) - y^\delta\|^p &= \frac{1}{p}\|F(x^\dagger) - \gamma\,G\,\omega + R_2 - y^\delta\|^p \\ &= \frac{\gamma^p}{p}\|G\,\omega\|^p + \gamma^{p-1}\langle J_p(G\,\omega), y^\delta - F(x^\dagger) - R_2\rangle \\ &\quad + D_{Y,p}(y^\delta - F(x^\dagger - \gamma\,\omega), \gamma\,G\,\omega). \end{aligned}$$

We now set $\gamma^{p-1} = \alpha$. We conclude by the minimizing property of x_α^δ

$$\frac{1}{p}\|F(x_\alpha^\delta) - y^\delta\|^p + \alpha\,\big(P(x_\alpha^\delta) - P(x^\dagger - \gamma\,\omega)\big) \le \frac{1}{p}\|F(x^\dagger - \gamma\,\omega) - y^\delta\|^p.$$

Hence, we can derive

$$\begin{aligned} \frac{1}{p}\|F(x_\alpha^\delta) - y^\delta\|^p + \alpha D_{\xi^\dagger}(x_\alpha^\delta, x^\dagger) &\le \alpha D_{\xi^\dagger}(x^\dagger - \gamma\,\omega, x^\dagger) + D_{Y,p}(y^\delta - F(x^\dagger - \gamma\,\omega), \gamma\,G\,\omega) \\ &\quad - \left(1 - \frac{1}{p}\right)\alpha\gamma\|G\,\omega\|^p - \alpha\langle J_p(G\,\omega), G(x_\alpha^\delta - x^\dagger)\rangle \\ &\quad + \alpha\langle J_p(G\,\omega), y^\delta - F(x^\dagger) - R_2\rangle \\ &\quad + \alpha\|v\|\,\big(\|x_\alpha^\delta - x^\dagger\| + \gamma\|\omega\|\big) \\ &= \alpha D_{\xi^\dagger}(x^\dagger - \gamma\,\omega, x^\dagger) + D_{Y,p}(y^\delta - F(x^\dagger - \gamma\,\omega), \gamma\,G\,\omega) \\ &\quad - \alpha\gamma\frac{p-1}{p}\|G\,\omega\|^p + \alpha\|v\|\,\big(\|x_\alpha^\delta - x^\dagger\| + \gamma\|\omega\|\big) \\ &\quad - \alpha\langle J_p(G\,\omega), y^\delta - F(x_\alpha^\delta) + R_1 - R_2\rangle. \end{aligned}$$

We apply Young's inequality (3.13) with $p_1 = p$, $a = \|F(x_\alpha^\delta) - y^\delta\|$ and $b = \alpha\|G\,\omega\|^{p-1}$. Then with $p_2 = \frac{p}{p-1}$ we obtain

$$\alpha\|G\,\omega\|^{p-1}\|F(x_\alpha^\delta) - y^\delta\| \le \frac{1}{p}\|F(x_\alpha^\delta) - y^\delta\|^p + \frac{p-1}{p}\alpha^{\frac{p}{p-1}}\|G\,\omega\|^p.$$

Since $\alpha^{\frac{p}{p-1}-1} = \alpha^{\frac{1}{p-1}} = \gamma$ we obtain $\alpha^{\frac{p}{p-1}} = \alpha\gamma$ and hence

$$\begin{aligned}\alpha D_{\xi^\dagger}(x_\alpha^\delta, x^\dagger) \;\le\; & \alpha D_{\xi^\dagger}(x^\dagger - \gamma\,\omega, x^\dagger) + D_{Y,p}(y^\delta - F(x^\dagger - \gamma\,\omega), \gamma\,G\,\omega) \\ & +\alpha\|v\|\left(\|x_\alpha^\delta - x^\dagger\| + \gamma\|\omega\|\right) \\ & \alpha\,L\,\|G\,\omega\|^{p-1}D_{\xi^\dagger}(x_\alpha^\delta, x^\dagger) + \alpha\,L\,\|G\,\omega\|^{p-1}D_{\xi^\dagger}(x^\dagger - \gamma\,\omega, x^\dagger)\end{aligned}$$

by using the estimates for $\|R_1\|$ and $\|R_2\|$. This proves the lemma. ■

Remark 3.3.1 *The main idea of the proof is to start with $T_{\alpha,y^\delta}(x_\alpha^\delta) \le T_{\alpha,y^\delta}(x^\dagger - \gamma\,\omega)$ instead of $x^\dagger$ as argument in the right side of the inequality as in the previous section. This choice based on the following observation: we consider classical Tikhonov regularization for linear equations in Hilbert spaces for exact data, i.e. we set $p = 2$, $\delta = 0$ and $P(x) := \frac{1}{2}\|x\|^2$. Then $\gamma = \alpha$ holds. Assume further $x^\dagger = A^*A\,\omega$ for some $\omega \in \mathcal{X}$. Then we derive*

$$x_\alpha := \left(A^*A + \alpha I\right)^{-1}\left(A^*y\right) = x^\dagger - \alpha\,\omega + c(\alpha)\tilde{\omega}$$

with $\tilde{\omega} \in \mathcal{X}$ and $\frac{c(\alpha)}{\alpha} \to 0$ as $\alpha \to 0$. So $x^\dagger - \alpha\,\omega$ seems to be a better approximation of x_α (and hence hopefully also for x_α^δ) than $x^\dagger$ which leads to the better convergence rate. This observation was already used in [58]. However, this idea is not applicable for nonlinear operators with degree (c_1, c_2) of nonlinearity with $c_2 > 0$. It is left open if we possibly can overcome this problem when the element $x^\dagger - \alpha\,\omega$ is replaced by another term.

For deriving error bounds we have to find bounds for the residuals $D_{\xi^\dagger}(x^\dagger - \gamma\,\omega, x^\dagger)$ and $D_{Y,p}(y^\delta - G(x^\dagger - \gamma\omega), \gamma\,G\,\omega)$. With condition (A8) we can estimate the term $D_{\xi^\dagger}(x^\dagger - \gamma\,\omega, x^\dagger)$ with

$$D_{\xi^\dagger}(x^\dagger - \gamma\,\omega, x^\dagger) \le G_r\|\gamma\,\omega\|^r = G_r\alpha^{\frac{r}{p-1}}\|\omega\|^r. \tag{3.32}$$

We now deal with estimates for the second term $D_{Y,p}(y^\delta - G(x^\dagger - \gamma\omega), \gamma\,G\,\omega)$. We first formulate a helpful lemma.

Lemma 3.3.2 *Suppose $p > 1$. Then under assumption (A9) there exists a constant $C_1 > 0$ such that*

$$D_{Y,p}(y_0 + \Delta y, y_0) \le C_1\|\Delta y\|^a \int_0^1 t^{a-1}\left(\max\{\|y_0 + t\,\Delta y\|, \|y_0\|\}\right)^{p-a} dt$$

holds for arbitrary $\Delta y, y_0 \in \mathcal{Y}$. Here, the constant C_1 depends only on p and $\mathcal{Y}$.

PROOF. We set $H(t) := \max\{\|y_0 + t\,\Delta y\|, \|y_0\|\}$. Then from (A8) and the Xu/Roach inequalities, see Proposition 2.2.2(ii), we obtain

$$D_{Y,p}(y_0 + \Delta y, y_0) \leq C_1 \int_0^1 \frac{H(t)^p}{t} \left(\frac{t\,\|\Delta y\|}{H(t)} \right)^a dt.$$

This immediately shows the assertion. ■

Now we can prove the following estimate.

Lemma 3.3.3 *Suppose $p > 1$. Let (A1)-(A6),(A8),(A9) and (3.28) hold. Assume further that $\delta \leq \tilde{c}\gamma$ for some constant $\tilde{c} > 0$ if $p > a$. If $x^\dagger - \gamma\omega \in \mathcal{D}$ then there exist two constants $C_2 > 0$ and $C_3 > 0$ such that*

$$D_{Y,p}(y^\delta - F(x^\dagger - \gamma\omega), \gamma\, G\,\omega) \leq \alpha \left(C_2 \alpha^{\frac{r}{p-1}} + C_3 \delta^a \alpha^{\frac{1-a}{p-1}} \right). \tag{3.33}$$

PROOF. We recall the notation $R_2 := F(x^\dagger - \gamma\omega) - F(x^\dagger) + \gamma\, G\,\omega$. Then we have

$$\|R_2\| \leq L\, D_{\xi^\dagger}(x^\dagger - \gamma\,\omega, x^\dagger) \leq L\, G_r \|\omega\|^r \alpha^{\frac{r}{p-1}}.$$

We apply Lemma 3.3.2 with $y_0 := \gamma\, G\,\omega$ and $\Delta y := y^\delta - F(x^\dagger - \gamma\omega) - \gamma\, G\,\omega$. Then we estimate

$$\begin{aligned} \|y^\delta - F(x^\dagger - \gamma\omega) - \gamma\, G\,\omega\|^a &= \|y^\delta - F(x^\dagger) - R_2\|^a \\ &\leq (\delta + \|R_2\|)^a \\ &\leq 2^{a-1} \left(\delta^a + (L\, G_r)^a \|\omega\|^{r\,a} \alpha^{\frac{r\,a}{p-1}} \right). \end{aligned}$$

We now consider $H(t) := \max\left\{\|t(y^\delta - F(x^\dagger - \gamma\omega) - \gamma\, G\,\omega) + \gamma\, G\,\omega\|, \gamma\|G\,\omega\|\right\}$. Then we apply the estimate

$$H(t) \geq \gamma\|G\,\omega\| = \alpha^{\frac{1}{p-1}} \|G\,\omega\|$$

for $p < a$ and

$$\begin{aligned} H(t) &\leq \|y^\delta - F(x^\dagger - \gamma\omega) - \gamma\, G\,\omega\| + \gamma\|G\,\omega\| \\ &\leq \delta + \|R_2\| + \|G\,\omega\|\alpha^{\frac{1}{p-1}} \\ &\leq \delta + L\,G_r\|\omega\|^r\alpha^{\frac{r}{p-1}} + \|G\,\omega\|\alpha^{\frac{1}{p-1}} \\ &\leq \left(\tilde{c} + L\, G_r \|\omega\|^r \alpha_{max}^{\frac{r-1}{p-1}} + \|G\,\omega\| \right) \alpha^{\frac{1}{p-1}} \end{aligned}$$

for $p > a$. We set $\mathcal{C}_1 := 2^{a-1}(L\, G_r)^a \|\omega\|^{r\,a}$ and $\mathcal{C}_2 := \|G\,\omega\|^{p-a}$ if $p \leq a$ and

$$\mathcal{C}_2 := \left(\tilde{c} + L\, G_r \|\omega\|^r \alpha_{max}^{\frac{r-1}{p-1}} + \|G\,\omega\| \right)^{p-a}$$

if $p > a$. Then we derive

$$\begin{aligned} D_{Y,p}(y^\delta - F(x^\dagger - \gamma\omega), \gamma G\omega) &\le \frac{C_1}{a}\left(2^{a-1}\delta^a + C_1\alpha^{\frac{r\,a}{p-1}}\right) C_2\alpha^{\frac{p-a}{p-1}} \\ &\le \alpha\frac{2^{a-1}C_1C_2}{a}\delta^a\alpha^{\frac{1-a}{p-1}} + \frac{C_1C_1C_2}{a}\alpha^{\frac{r\,a-a+1}{p-1}} \\ &\le \alpha\frac{2^{a-1}C_1C_2}{a}\delta^a\alpha^{\frac{1-a}{p-1}} + \frac{C_1C_1C_2}{a}\alpha_{max}^{\frac{(r-1)(a-1)}{p-1}}\alpha^{\frac{r}{p-1}}. \end{aligned}$$

In the last estimate we used that

$$r\,a - a + 1 = r + (r\,a - a - r + 1) = r + (r-1)(a-1).$$

This proves the lemma. ■

We present a first convergence rate result, provided the element $\xi^\dagger \in \mathcal{X}^*$ satisfies the source condition (3.29), see [35, Theorem 6.2].

Theorem 3.3.1 (Convergence rates VI) *Suppose $p > 1$. Assume (A1)-(A6),(A8), (A9), (3.28) and $\xi^\dagger \in \mathcal{X}^*$ satisfies the source condition (3.29) for some $\omega \in \mathcal{X}$. If $L\,\|G\,\omega\|^{p-1} < 1$ and there exists $\bar\gamma > 0$ such that $x^\dagger - \gamma\,\omega \in \mathcal{D}$ for all $\gamma \in (0, \bar\gamma]$ then the parameter choice $\alpha \sim \delta^{\frac{a(p-1)}{r+a-1}}$ leads to the convergence rate*

$$D_{\xi^\dagger}(x_\alpha^\delta, x^\dagger) = \mathcal{O}\left(\delta^{\frac{a\,r}{r+a-1}}\right) \quad \textit{as} \quad \delta \to 0. \tag{3.34}$$

PROOF. We now have the estimate

$$D_{\xi^\dagger}(x_\alpha^\delta, x^\dagger) \le \frac{C_3}{1 - L\,\|G\,\omega\|^{p-1}}\delta^a\alpha^{\frac{1-a}{p-1}} + \frac{(1 + L\,\|G\,\omega\|^{p-1})\,G_r\|\omega\|^r + C_2}{1 - L\,\|G\,\omega\|^{p-1}}\alpha^{\frac{r}{p-1}}.$$

The usual balancing gives

$$\alpha^{\frac{r}{p-1}+\frac{a-1}{p-1}} = \alpha^{\frac{r+a-1}{p-1}} = \delta^a \;\Leftrightarrow\; \alpha = \delta^{\frac{a\,(p-1)}{r+a-1}}$$

which proves the according parameter choice and the convergence rate. We only have to observe that

$$\frac{\delta}{\alpha^{\frac{1}{p-1}}} \sim \delta^{1-\frac{a}{r+a-1}} = \delta^{\frac{r-1}{r+a-1}} \to 0$$

as $\delta \to 0$. In particular, the additional assumption in Lemma 3.3.2 for $p > a$ can be fulfilled uniformly for arbitrary $\delta \le \delta_{max}$. ■

We see that the smoothness of the penalty functional P and the space $\mathcal{Y}$ influence the achieved convergence rates. It turns out, that for $1 < r, a < 2$ we achieve weaker rates than for $a = r = 2$, which is devoted the weaker smoothness of the Tikhonov functional. This is opposite to dealing with convergence rates for the weaker reference source condition (3.11) in the previous section, where no assumptions on the smoothness of P and $\mathcal{Y}$ were needed. Moreover, the parameter p does not influence the convergence rate (3.34). Only the corresponding parameter choice $\alpha = \alpha(\delta)$ depends on this parameter. We consider some specific situations in the following example.

Example 3.3.3 *We apply Theorem 3.3.1 to the following situations:*

(a) *If* $r = 2$ *then* $\alpha = \delta^{\frac{a(p-1)}{a+1}}$ *and* $D_{\xi^\dagger}(x_\alpha^\delta, x^\dagger) \sim \delta^{\frac{2a}{a+1}}$ *as* $\delta \to 0$. *In particular, we observe that* $\frac{2a}{a+1} \to 1$ *for* $a \to 1$ *and* $\frac{2a}{a+1} \to \frac{4}{3}$ *for* $a \to 2$.

(b) *If* $a = 2$ *then* $\alpha = \delta^{\frac{2(p-1)}{r+1}}$ *and* $D_{\xi^\dagger}(x_\alpha^\delta, x^\dagger) \sim \delta^{\frac{2r}{r+1}}$ *as* $\delta \to 0$.

(c) *In the classical situation* $r = a = p = 2$ *we have* $\alpha = \delta^{\frac{2}{3}}$ *and* $D_{\xi^\dagger}(x_\alpha^\delta, x^\dagger) \sim \delta^{\frac{4}{3}}$ *as* $\delta \to 0$. *If* $\mathcal{X}$ *is a Hilbert space and* $P(x) = \frac{1}{2}\|x - x_*\|^2$ *for some* $x_* \in \mathcal{X}$ *then we have* $D_{\xi^\dagger}(x_\alpha^\delta, x^\dagger) = \frac{1}{2}\|x_\alpha^\delta - x^\dagger\|^2$ *and hence* $\|x_\alpha^\delta - x^\dagger\| \sim \delta^{\frac{2}{3}}$ *which is the expected optimal rate for Tikhonov regularization in Hilbert spaces.*

Now we consider the case $p = 1$ which need some special treatment. As opposite to $p > 1$ we introduce two changes in the notation. First, we assume $y_2 \neq 0$ and define the Bregman distance

$$D_{Y,1}(y_1, y_2) := \|y_1\| - \|y_2\| - \frac{1}{\|y_2\|^{a-1}} \langle J_a(y_2), y_1 - y_2 \rangle, \qquad y_1, y_2 \in \mathcal{Y}.$$

Here we used that

$$\lim_{\varepsilon \to 0} \frac{\|y_2 + \varepsilon\,(y_1 - y_2)\| - \|y_2\|}{\varepsilon} = \frac{1}{\|y_2\|^{a-1}} \langle J_a(y_2), y_1 - y_2 \rangle$$

holds for all $y_1, y_2 \in \mathcal{Y}$, $y_2 \neq 0$. In a second modification we assume the source condition

$$\xi^\dagger = G^\star J_a(G\,\omega), \qquad \omega \in \mathcal{X}. \tag{3.35}$$

In particular, here the smoothness a of the space $\mathcal{Y}$ now describes the choice of the accordant duality mapping J_a in the source condition. We give an first estimate concerning the Bregman distance $D_{Y,1}(y_1, y_2)$.

Lemma 3.3.4 *Under assumption (A9) there exists a constant* $C_4 > 0$ *such that*

$$D_{Y,1}(y_0 + \Delta y, y_0) \leq C_4 \frac{\|\Delta y\|^a}{\|y_0\|^{a-1}}$$

holds for all $y_0, \Delta y \in \mathcal{Y}$.

PROOF. From the mean value theorem we have

$$\begin{aligned}
D_{Y,1}(y_0 + \Delta y, y_0) &= \|y_0 + \Delta y\| - \|y_0\| - \frac{1}{\|y_0\|^{a-1}} \langle J_a(y_0), \Delta y\rangle \\
&= \int_0^1 \frac{1}{\|y_0 + t\,\Delta y\|^{a-1}} \langle J_a(y_0 + t\,\Delta y), \Delta y\rangle - \frac{1}{\|y_0\|^{a-1}} \langle J_a(y_0), \Delta y\rangle\, dt \\
&= \int_0^1 \frac{1}{\|y_0\|^{a-1}} \langle J_a(y_0 + t\,\Delta y) - J_a(y_0), \Delta y\rangle\, dt \\
&\quad + \int_0^1 \frac{\|y_0\|^{a-1} - \|y_0 + t\,\Delta y\|^{a-1}}{\|y_0\|^{a-1}\|y_0 + t\,\Delta y\|^{a-1}} \langle J_a(y_0 + t\,\Delta y), \Delta y\rangle\, dt \\
&\leq \frac{\|\Delta y\|}{\|y_0\|^{a-1}} \int_0^1 \|J_a(y_0 + t\,\Delta y) - J_a(\Delta y)\|\, dt \\
&\quad + \frac{\|\Delta y\|}{\|y_0\|^{a-1}} \int_0^1 \left|\|y_0 + t\,\Delta y\|^{a-1} - \|y_0\|^{a-1}\right|\, dt.
\end{aligned}$$

For the estimate of the second integral we used that $\|J_a(y_0 + t\,\Delta y)\| = \|y_0 + t\,\Delta y\|^{a-1}$. For the first integral we now apply [96, Remark 4], i.e. the estimate

$$\|J_a(y_0 + t\,\Delta y) - J_a(y_0)\| \leq C\,\|t\,\Delta y\|^{a-1} = C\,\|\Delta y\|^{a-1} t^{a-1}$$

holds for some $C > 0$. For the second integral we apply that $0 < a - 1 \leq 1$ with gives

$$\|y_0 + t\,\Delta y\|^{a-1} \leq \|y_0\|^{a-1} + t^{a-1}\|\Delta y\|^{a-1}$$

which we use if $\|y_0 + t\,\Delta y\| \geq \|y_0\|$ and

$$\|y_0\|^{a-1} = \|y_0 + t\,\Delta y - t\,\Delta y\|^{a-1} \leq \|y_0 + t\,\Delta y\|^{a-1} + t^{a-1}\|\Delta y\|^{a-1}$$

which will applied if $\|y_0\| \geq \|y_0 + t\,\Delta y\|$. Hence we end at

$$D_{Y,1}(y_0 + \Delta y, y_0) \leq \frac{C+1}{a} \frac{\|\Delta y\|^a}{\|y_0\|^{a-1}},$$

which proves the lemma. ■

We can present the following convergence rates result, see also [36, Theorem 6.3].

Theorem 3.3.2 (Convergence rates VII) *Suppose $p = 1$. Assume (A1)-(A6),(A8), (A9), (3.28), $\xi^\dagger \in \mathcal{X}^*$ satisfies the source condition (3.35) for some $\omega \in \mathcal{X}$ with $L\,\|G\,\omega\| < 1$ and there exists $\bar{\gamma} > 0$ such that $x^\dagger - \gamma\,\omega \in \mathcal{D}$ for all $\gamma \in (0, \bar{\gamma}]$. If the regularization parameter $\alpha > 0$ is chosen such that $0 \leq 1 - \alpha\|G\,\omega\|^{a-1} \leq \delta^{\frac{a(r-1)}{r+a-1}}$ then the convergence rate (3.34) holds.*

PROOF. Following the calculations in the proof of Lemma 3.3.1 we derive

$$\begin{aligned}\|F(x_\alpha^\delta) - y^\delta\| + \alpha D_{\xi^\dagger}(x_\alpha^\delta, x^\dagger) \leq\, & \alpha D_{\xi^\dagger}(x^\dagger - \gamma\omega, x^\dagger) + D_{Y,1}(y^\delta - F(x^\dagger - \gamma\omega), \gamma G\,\omega) \\ & + \alpha\langle J_a(G\,\omega), G\,(x^\dagger - x_\alpha^\delta)\rangle - \alpha\gamma\|G\,\omega\|^a \\ & + \gamma\|G\,\omega\| + \frac{1}{\|G\,\omega\|^{a-1}}\langle J_a(G\,\omega), y^\delta - R_2 - F(x^\dagger)\rangle \\ \leq\, & \alpha D_{\xi^\dagger}(x^\dagger - \gamma\omega, x^\dagger) + D_{Y,1}(y^\delta - F(x^\dagger - \gamma\omega), \gamma G\,\omega) \\ & + \alpha\langle J_a(G\,\omega), F(x^\dagger) - F(x_\alpha^\delta) + R_1\rangle - \alpha\gamma\|G\,\omega\|^a \\ & + \gamma\|G\,\omega\| + \frac{1}{\|G\,\omega\|^{a-1}}\langle J_a(G\,\omega), y^\delta - F(x^\dagger)\rangle \\ & + L\, D_{\xi^\dagger}(x^\dagger - \gamma\,\omega, x^\dagger) \\ \leq\, & (\alpha + L) D_{\xi^\dagger}(x^\dagger - \gamma\omega, x^\dagger) + D_{Y,1}(y^\delta - F(x^\dagger - \gamma\omega), \gamma G\,\omega) \\ & + \alpha\|G\,\omega\|^{a-1}\|y^\delta - F(x_\alpha^\delta)\| + \alpha\, L\,\|G\,\omega\|^{a-1} D_{\xi^\dagger}(x_\alpha^\delta, x^\dagger) \\ & + \gamma\|G\,\omega\|(1 - \alpha\|G\,\omega\|^{a-1}) + \left|1 - \alpha\|G\,\omega\|^{a-1}\right|\delta.\end{aligned}$$

Since $\alpha\|G\,\omega\|^{a-1} \leq 1$ we have

$$\begin{aligned}\alpha\left(1 - L\,\|G\,\omega\|^{a-1}\right) D_{\xi^\dagger}(x_\alpha^\delta, x^\dagger) \leq\, & (\alpha + L) D_{\xi^\dagger}(x^\dagger - \gamma\omega, x^\dagger) + D_{Y,1}(y^\delta - F(x^\dagger - \gamma\omega), \gamma G\,\omega) \\ & + \gamma\|G\,\omega\|(1 - \alpha\|G\,\omega\|^{a-1}) + \left|1 - \alpha\|G\,\omega\|^{a-1}\right|\delta.\end{aligned}$$

Again we use $D_{\xi^\dagger}(x^\dagger - \gamma\omega, x^\dagger) \leq G_r\|\omega\|^r\gamma^r$. We apply the above lemma with $y_0 := \gamma\,G\,\omega$ and $\Delta y := y^\delta - F(x^\dagger - \gamma\omega) - \gamma\,G\,\omega$ and obtain as in the last section

$$\begin{aligned}D_{Y,1}(y^\delta - F(x^\dagger - \gamma\omega), \gamma\,G\,\omega) \;&\leq\; C_4\frac{\|y^\delta - F(x^\dagger - \gamma\omega) - \gamma\,G\,\omega\|^a}{\gamma^{a-1}\|G\,\omega\|^{a-1}} \\ &\leq\; C_4\frac{(\delta + L\,G_r\|\omega\|^r\gamma^r)^a}{\gamma^{a-1}\|G\,\omega\|^{a-1}}.\end{aligned}$$

We now set $\gamma := \delta^{\frac{a}{r+a-1}}$. Since $1 - \alpha\|G\,\omega\|^{a-1} \leq \delta^{\frac{a(r-1)}{r+a-1}}$ we derive

$$\delta^a\gamma^{1-a} = \delta^{a - \frac{a(a-1)}{r+a-1}} = \delta^{\frac{r\,a}{r+a-1}}$$

and

$$\gamma^{r\,a-a+1} = \delta^{\frac{r\,a^2 - a^2 + a}{r+a-1}} = \delta^{\frac{r\,a}{r+a-1} + \frac{r\,a(a-1) - a^2 + a}{r+a-1}} = \delta^{\frac{r\,a}{r+a-1}}\delta^{\frac{(r-1)(a-1)a}{r+a-1}}.$$

Moreover, we can estimate

$$\gamma\left(1 - \alpha\|G\,\omega\|^{a-1}\right) \leq \delta^{\frac{r\,a}{r+a-1}}$$

and

$$\delta\left(1 - \alpha\|G\,\omega\|^{a-1}\right) \leq \delta^{\frac{r\,a}{r+a-1} + \frac{r+a-1-a}{r+a-1}} = \delta^{\frac{r\,a}{r+a-1}}\delta^{\frac{r-1}{r+a-1}}.$$

This proves the accordant convergence rate. ■

Here again as in the situation of low order convergence rates for the case $p = 1$ the optimal choice of the regularization parameter α depends essentially on $\|G\,\omega\|$. Moreover, this dependency is the reason, that presenting convergence rates in terms of approximate source conditions does not seem to be promising for $p = 1$. For a given choice $R = R(\alpha)$, $\|\omega\| \leq R$, we have to ensure the condition $0 \leq 1 - \alpha\|G\,\omega\|^{a-1} \leq \delta^{\frac{a(r-1)}{r+a-1}}$ which does not seem to be reasonable under general conditions. Therefore we skip the case $p = 1$ in the further considerations.

3.3.3 Approximate source conditions again

For violated source condition (3.29) we consider only linear equations since a smallness assumption was applied for proving Lemma 3.3.1 for nonlinear equations. Then for proving convergence rates with aid of approximate source conditions we apply the idea of distance functions again. As opposite to Section 3.2 the reference source condition is now given by (3.29). We present the accordant definition.

Definition 3.3.1 *For given $\xi \in \mathcal{X}^*$ and $1 < p < \infty$ the distance function $d_p(\cdot;\xi) : [0,\infty) \longrightarrow \mathbb{R}$ is defined as*

$$d_p(R;\xi) := \inf\left\{\|\xi - A^\star J_p(A\,\omega)\| \;:\; \omega \in \mathcal{X},\ \|\omega\| \leq R^{\frac{1}{p-1}}\right\}, \qquad R \geq 0.$$

In a first step we show, that we can in fact replace the infimum by the minimum in the definition, i.e. for arbitrary $\xi^\dagger \in \mathcal{X}^*$ and each $R \geq 0$ we can find a element $\omega = \omega(R)$ such that $d_p(R;\xi) = \|\xi - A^\star J_p(A\,\omega(R))\|$.

Lemma 3.3.5 *For arbitrary $\xi \in \mathcal{X}^*$ the minimizing problem*

$$\|\xi - A^\star J_p(A\,\omega)\| \to \min \quad \textit{subject to} \quad \|\omega\| \leq R^{\frac{1}{p-1}}$$

admits a solution for each $R \geq 0$.

PROOF. By [98, Theorem 38.A], we have to show that the functional $\omega \mapsto \|\xi - A^\star J_p(A\,\omega)\|$ is weakly lower semi-continuous. By [98, Proposition 38.7], this is equivalent to: for all $\varepsilon > 0$ there exists $\delta(\varepsilon) > 0$ such that

$$\|\omega - \tilde{\omega}\| < \delta(\varepsilon) \;\Rightarrow\; \|\xi - A^\star J_p(A\,\omega)\| < \|\xi - A^\star J_p(A\,\tilde{\omega})\| + \varepsilon.$$

For simplicity, we assume $\varepsilon < \|A\|$. We know, that J_p is uniformly continuous on each bounded set, see [98, Proposition 47.19(2(ii))]. Hence there exists a $\tilde{\delta}(\varepsilon\|A\|^{-1})$ such that $\|\omega - \tilde{\omega}\| < \tilde{\delta}(\varepsilon\|A\|^{-1})$ implies $\|J_p(\omega) - J_p(\tilde{\omega})\| < \varepsilon\|A\|^{-1}$ for all $\tilde{\omega} \in \mathcal{B}_1(\omega)$. Hence

$$\begin{aligned}\|\xi - A^\star J_p(A\,\omega)\| &= \|\xi - A^\star J_p(A\,\tilde{\omega}) - A^\star J_p(A\,\omega) + A^\star J_p(A\,\tilde{\omega})\| \\ &\leq \|\xi - A^\star J_p(A\,\tilde{\omega})\| + \|A\|\,\|J_p(A\,\omega) - J_p(A\,\tilde{\omega})\|.\end{aligned}$$

We set $\delta(\varepsilon) := \tilde{\delta}(\varepsilon\|A\|^{-1})\|A\|^{-1}$. Then $\|\omega - \tilde{\omega}\| < \delta(\varepsilon)$ implies

$$\|A\,\omega - A\,\tilde{\omega}\| \leq \|A\|\,\|\omega - \tilde{\omega}\| < \tilde{\delta}(\varepsilon\|A\|^{-1})$$

and hence $\|A\|\,\|J_p(A\,\omega) - J_p(A\,\tilde{\omega})\| < \varepsilon$. This proves the assertion. ■

We first motivate the specific definition of the distance functions $d_p(R;\xi)$. The underlying results is formulated in the following lemma, see also [37].

Lemma 3.3.6 *There exists two constants $\kappa_1, \kappa_2 > 0$ such that the inequalities*

$$d_2(\kappa_1^{p-2}R;\xi) \le d_p(R;\xi) \le d_2(\kappa_2^{p-2}R;\xi)$$

hold for $p < 2$ and the estimates

$$d_2(\kappa_2^{p-2}R;\xi) \le d_p(R;\xi) \le d_2(\kappa_1^{p-2}R;\xi)$$

are valid for $p > 2$.

PROOF. First we set $\tilde{\omega} := R^{-\frac{1}{p-1}}\omega$. Hence, for $\|\omega\| \le R^{\frac{1}{p-1}}$ we conclude $\|\tilde{\omega}\| \le 1$ and

$$\begin{aligned}\|\xi - A^\star J_p(A\,\omega)\| = \left\|\xi - A^\star J_p\left(R^{\frac{1}{p-1}}A\,\tilde{\omega}\right)\right\| &= \|\xi - R\,A^\star J_p(A\,\tilde{\omega})\| \\ &= R\left\|\frac{1}{R}\xi - A^\star J_p(A\,\tilde{\omega})\right\|.\end{aligned}$$

So we can rewrite the definition of $d_p(R;\xi)$ as

$$d_p(R;\xi) := R\,\inf\left\{\left\|\frac{1}{R}\xi - A^\star J_p(A\,\tilde{\omega})\right\| \;:\; \tilde{\omega} \in \mathcal{X},\, \|\tilde{\omega}\| \le 1\right\}. \tag{3.36}$$

Furthermore, with the relation $J_p(A\,\tilde{\omega}) = \|A\,\tilde{\omega}\|^{p-2}J_2(A\,\tilde{\omega})$

$$R\left\|\frac{1}{R}\xi - A^\star J_p(A\,\tilde{\omega})\right\| = \left(R\|A\,\tilde{\omega}\|^{p-2}\right)\left\|\frac{1}{R\|A\,\tilde{\omega}\|^{p-2}}\xi - A^\star J_2(A\,\tilde{\omega})\right\|$$

yields. This implies $d_p(R;\xi) = d_2(R\|A\,\tilde{\omega}(R)\|^{p-2};\xi)$ where $\tilde{\omega}(R)$ satisfies $d_p(R;\xi) = R\|R^{-1}\xi - A^\star J_p(A\,\tilde{\omega}(R))\|$. We need estimates for the term $\|A\,\tilde{\omega}\|$ from above and below. Clearly $\|A\,\tilde{\omega}\| \le \|A\|$ since $\|\tilde{\omega}\| \le 1$. For a lower bound we set $c(R) := d_\mu(R;\xi)\|\xi\|^{-1}$ for given $\mu > 2$ and $\xi \ne 0$. We conclude

$$\|\xi\|(1 - c(R)) \le \|A^\star J_\mu(A\,\omega(R))\| = \|A\|\,\|J_\mu(A\,\omega(R))\| = \|A\|\,R\,\|A\,\tilde{\omega}(R)\|^{\mu-1},$$

which shows

$$\|A\,\tilde{\omega}\| \ge \left(\frac{\|\xi\|}{\|A\|}\right)^{\frac{1}{\mu-1}}\left(\frac{1-c(R)}{R}\right)^{\frac{1}{\mu-1}} \ge \min\left\{1, \left(\frac{\|\xi\|}{\|A\|}\right)\right\}\left(\frac{1-c(R)}{R}\right)^{\frac{1}{\mu-1}}.$$

The estimate holds for all $\mu > 2$. Hence the term $[(1 - c(R))R^{-1}]^{1/(\mu-1)}$ can be chosen arbitrary close to one by choosing the parameter μ arbitrary large. We set $\kappa_1 := \min\{1, \|\xi\|\|A\|^{-1}\}$, $\kappa_2 := \|A\|$ and conclude $\kappa_1 \le \|A\,\tilde{\omega}\| \le \kappa_2$. By the monotonicity of the distance functions we have proved the lemma. ■

This shows that the influence of the parameter p on the decay rate of the distance functions is rather low. If, in particular, we can suppose power-type distance functions, i.e. $d_{p_i}(R;\xi) = C_i R^{-\mu_i}$, $C_i, \mu_i > 0$, $i = 1, 2$, for two parameters $p_1, p_2 > 1$ then the calculations above show $\mu_1 = \mu_2$.

To shorten the notation we introduce the set $\mathcal{M} := \mathcal{R}(A^\star J_p(A\,\cdot))$. We now are able to present convergence rates, if the source condition (3.29) is not satisfied for any $R > 0$, i.e. $\xi^\dagger \notin \mathcal{M}$, but $\xi^\dagger \in \overline{\mathcal{M}}$. This additional assumption seems to be quite restrictive at first view. On the other hand, the following statement yields, see [37, Lemma 4.1].

Lemma 3.3.7 *Assume the operators A and $A^\star$ to be injective. Then $\overline{\mathcal{M}} = \mathcal{X}^*$.*

Proof. By [94, Satz III.4.5] we have $\overline{\mathcal{R}(A^\star)} = \mathcal{X}^*$ and $\overline{\mathcal{R}(A)} = \mathcal{Y}$. Assume $\xi \notin \mathcal{M}$. On the other hand, for each $\varepsilon > 0$ there exists an element $\tilde{y}_\varepsilon \in \mathcal{Y}^*$ with $\|\xi - A^\star y_\varepsilon\| < \varepsilon$. Since J_p is bijective, there exists $y_\varepsilon \in \mathcal{Y}$ with $J_p(y_\varepsilon) = \tilde{y}_\varepsilon$. Finally, there exists $\omega_\varepsilon \in \mathcal{X}$ with $\|y_\varepsilon - A\,\omega_\varepsilon\| < \varepsilon$. Hence,

$$\begin{aligned} \|\xi - A^\star J_p(A\,\omega_\varepsilon)\| &\le \|\xi - A^\star J_p(y_\varepsilon)\| + \|A^\star\|\,\|J_p(A\,\omega_\varepsilon) - J_p(y_\varepsilon)\| \\ &< \varepsilon + \|A^\star\|\,\|J_p(A\,\omega_\varepsilon) - J_p(y_\varepsilon)\| \end{aligned}$$

Since J_p is continuously we conclude $\|J_p(A\,\omega_\varepsilon) - J_p(y_\varepsilon)\| \to 0$ for $\varepsilon \to 0$ which implies $A^\star J_p(\omega_\varepsilon) \to \xi$ for $\varepsilon \to 0$. Consequently, $\xi \in \overline{\mathcal{M}}$ yields. ■

As in in the case of low order convergence rates we present an example for an upper bound of such distance function.

Example 3.3.4 *We return to our Example 1.2.1 with arbitrary $1 < p, q < \infty$. Moreover we choose $P(x) := \frac{1}{q}\|x\|^q$. We again suppose $x^\dagger \equiv 1$. Then $P'(x^\dagger) = J_q(x^\dagger) \equiv 1$ and the source condition (3.29) is obviously violated. We going on to find a majorant for the distance function $d_p(R; J_q(x^\dagger))$, $R \ge 0$. Therefore we define the functions*

$$\omega_R(t) := \begin{cases} 0, & 0 \le t < 1 - c_2 R^{-\mu}, \\ c_1 R^{\frac{\mu p}{p-1}}, & 1 \ge t \ge 1 - c_2 R^{-\mu}, \end{cases}$$

for $R>0$ sufficiently large, i.e. $1 - c_2 R^{-\mu} > 0$. Here we set $\mu := \frac{q}{p(q-1)+1}$ and the constants $c_1, c_2 > 0$ are chosen such that $c_1 = c_2^{-\frac{1}{q}}$ and $c_1^{p-1} c_2^{p} = p$ which leads to $c_2 := p^\mu$. Then we have

$$\|\omega_R\|_{L^q} = c_1 R^{\frac{\mu p}{p-1}} \left[\int_{1-c_2R^{-\mu}}^{1} 1\, dt \right]^{\frac{1}{q}} = c_1 c_2^{\frac{1}{q}} R^{\frac{\mu p}{p-1} - \frac{\mu}{q}} = R^{\frac{1}{p-1}}.$$

Furthermore we calculate

$$A\,\omega_R(t) := \int_0^t \omega_R(\tau)\, d\tau = \begin{cases} 0, & 0 \le t < 1 - c_2 R^{-\mu}, \\ c_1 R^{\frac{\mu p}{p-1}} \left(t - (1 - c_2 R^{-\mu})\right), & 1 \ge t \ge 1 - c_2 R^{-\mu}, \end{cases}$$

and

$$[J_p(A\,\omega_R)](t) = [(A\,\omega_R)(t)]^{p-1} = \begin{cases} 0, & 0 \le t < 1 - c_2 R^{-\mu}, \\ c_1^{p-1} R^{\mu p} \left(t - (1 - c_2 R^{-\mu})\right)^{p-1}, & 1 \ge t \ge 1 - c_2 R^{-\mu}. \end{cases}$$

An application of the adjoint operator $A^\star$ gives

$$\begin{aligned} [A^\star J_p(A\,\omega_R)](t) &:= \int_t^1 J_p(A\,\omega_R)(\tau)\, d\tau \\ &= \begin{cases} 1, & 0 \le t < 1 - c_2 R^{-\mu}, \\ 1 - \frac{c_1^{p-1}}{p} R^{\mu p} \left(t - (1 - c_2 R^{-\mu})\right)^{p}, & 1 \ge t \ge 1 - c_2 R^{-\mu}. \end{cases} \end{aligned}$$

Hence we arrive at

$$\begin{aligned}\|J_q(x^\dagger) - A^\star J_p(A\,\omega_R)\|_{L^{\frac{q}{q-1}}} &= \frac{c_1^{p-1}}{p} R^{\mu p}\left(\frac{q-1}{pq+q-1} c_2^{\frac{(p+1)q}{q-1}} R^{-\mu\left(\frac{pq}{q-1}+1\right)}\right)^{\frac{q-1}{q}} \\ &= c_2\left(\frac{q-1}{pq+q-1}\right)^{\frac{q-1}{q}} R^{-\frac{\mu(q-1)}{q}} = C\,R^{-\frac{q-1}{p(q-1)+1}}.\end{aligned}$$

Here we derive a power-type majorant for the distance function $d_p(R; J_q(x^\dagger))$. Moreover, in the Hilbert space setting $q = p = 2$ we derive $d_2(R; x^\dagger) \le C\,R^{-\frac{1}{3}}$. On the other hand, we already mentioned in Example 2.4.3 that $x^\dagger \in \mathcal{R}((A^*A)^\nu)$ for $0 < \nu < \frac{1}{4}$. From [17, Theorem 3.2], see also Example 2.4.1, with $\nu = \frac{1}{4}$ we conclude an estimate for the distance function of the same order $R^{-\frac{1}{3}}$. So in this situation the calculations of the example leads to an estimate for the distance function which seems of correct order.

We will also treat an example of an distance function when $\xi^\dagger \in \mathcal{X}^*$ violates the source condition (3.29) but satisfies the weaker source condition (3.11).

Example 3.3.5 *We consider the same situation as in the previous example but now assuming $\xi^\dagger(t) := 1-t$, $t \in (0,1)$. Hence $x^\dagger(t) := (1-t)^{\frac{1}{q-1}}$ is the underlying exact solution of equation (3.3). Here we set*

$$\omega_R(t) := \begin{cases} \frac{1}{c}, & 0 \le t \le c, \\ 0, & c < t \le 1, \end{cases}$$

for some $0 < c < 1$. This gives

$$R^{\frac{1}{p-1}} = \|\omega_R\|_{L^q} = c^{\frac{1}{q}} c^{-1} = c^{\frac{1-q}{q}} \;\Leftrightarrow\; c := R^{-\frac{q}{(p-1)(q-1)}}.$$

We further calculate

$$A\,\omega_R(t) := \begin{cases} \frac{t}{c}, & 0 \le t \le c, \\ 1, & c < t \le 1, \end{cases} \quad \text{and} \quad [J_p(A\,\omega_R)](t) := \begin{cases} \left(\frac{t}{c}\right)^{p-1}, & 0 \le t \le c, \\ 1, & c < t \le 1. \end{cases}$$

So we end at

$$[A^\star(J_p(A\,\omega_R))](t) := \begin{cases} \frac{c}{p}\left(1 - \left(\frac{t}{c}\right)^p\right) + 1 - c, & 0 \le t \le c, \\ 1-t, & c < t \le 1. \end{cases}$$

We estimate the error. Here we get

$$\begin{aligned}\left\|\xi^\dagger - A^\star(J_p(A\,\omega_R)\right\|_{L^{q^*}} &\le \left(\int_0^c |t-c|^{q^*}\,dt\right)^{\frac{1}{q^*}} \\ &= \left(\frac{1}{q^*+1}\right)^{\frac{1}{q^*}} c^{\frac{q^*+1}{q^*}} \\ &= \left(\frac{1}{q^*+1}\right)^{\frac{1}{q^*}} R^{-\frac{2q-1}{(p-1)(q-1)}}.\end{aligned}$$

Hence with $C := (q^*+1)^{-1/q^*}$ we again derive with

$$d_p(R;\xi^\dagger) \le C\, R^{-\frac{2q-1}{(p-1)(q-1)}}, \qquad R > 1,$$

a power-type upper bound for the distance function $d_p(R;\xi^\dagger)$, $R \ge 1$. In particular, for $p = q = 2$ we obtain $d_p(R;\xi^\dagger) \le R^{-3}$, $R \ge 1$. The theoretical results of [17] now imply with $\mu = 1$ that $x^\dagger \in \mathcal{R}((A^*A)^{\tilde\nu})$ for all $0 < \tilde\nu < \nu$ with

$$-3 = \frac{\nu}{\nu - \mu} = \frac{\nu}{\nu - 1} \;\Leftrightarrow\; \nu = \frac{3}{4}.$$

On the other hand we conclude from $x \equiv 1 \in \mathcal{R}((A^*A)^{\hat\nu})$ for all $\hat\nu < \frac{1}{4}$ that $x^\dagger = A^*1 \in \mathcal{R}((A^*A)^{\hat\nu + \frac{1}{2}})$, $\hat\nu < \frac{1}{4}$. Hence we end at the same source condition as predicted by the derived bound of the distance function. This again also proves that we cannot improve the exponent of the distance function in this specific Hilbert space situation.

3.3.4 Error bounds and convergence rates

As seen in Section 3.3.2 the presented convergence analysis depends on the specific parameter a, r and p. Later on, also the parameter s from the convexity condition (A7) appears in the convergence rates results. In order to simplify the following considerations we connect the choice of the parameter p with the smoothness a of the space $\mathcal{Y}$, i.e. we suppose $a := \min\{p, 2\}$. We recall that in particular L^p-spaces are $\min\{p,2\}$-smooth for $1 < p < \infty$. Hence, in the following we only consider the (canonical) choice of a $\min\{p,2\}$-smooth space $\mathcal{Y}$.

Remark 3.3.2 *The results of Section 3.3.2 show that connecting the choice of p with the smoothness a of the space $\mathcal{Y}$ is not crucial. It is just employed to keep the subsequent analysis and the representation of the corresponding results more simple. If the parameter p is chosen independently on the smoothness we can proceed as in Lemma 3.3.3 by distinguishing the cases $p > a$ and $p \le a$. However, then the smoothness a of the space $\mathcal{Y}$ occurs as additional parameter in the error bounds and convergence rates results which is probably more confusing than it contains really new information.*

We are now able to formulate the accordant error bound result, see also [37, Corollary 3.1].

Lemma 3.3.8 *Assume (A1)-(A6), (A8),(A9) and $\xi^\dagger \in \mathcal{M}_H(R,d)$ for some $R, d \ge 0$.*

(i) If $1 < p \le 2$ then the estimate

$$D_{\xi^\dagger}(x_\alpha^\delta, x^\dagger) \le \tilde K_1 \frac{\delta^p}{\alpha} + G_r \alpha^{\frac{r}{p-1}} R^{\frac{r}{p-1}} + d\left(\varrho_x + \alpha^{\frac{1}{p-1}} R^{\frac{1}{p-1}}\right) \tag{3.37}$$

holds, where $\tilde K_1 > 0$ is a constant which does not depend on R, d, δ and α.

(ii) If $p > 2$ and $\delta \le \tilde{c}\,\alpha^{\frac{1}{p-1}} R^{\frac{1}{p-1}}$ for some constant $\tilde{c} > 0$, then the estimate

$$D_{\xi^\dagger}(x_\alpha^\delta, x^\dagger) \le K_1 \frac{\delta^2}{\alpha^{\frac{1}{p-1}}} R^{\frac{p-2}{p-1}} + G_r \alpha^{\frac{r}{p-1}} R^{\frac{r}{p-1}} + d\left(\varrho_x + \alpha^{\frac{1}{p-1}} R^{\frac{1}{p-1}}\right). \tag{3.38}$$

holds, where $K_1 > 0$ is a constant which does not depend on R, d, δ and α.

PROOF. We have $\xi^\dagger = A^\star J_p(A\,\omega) + v$ with $\|\omega\| \le R^{\frac{1}{p-1}}$ and $\|v\| \le d$. From Lemma 3.3.1 with $L = 0$ and (3.32) we can start with

$$\begin{aligned} D_{\xi^\dagger}(x_\alpha^\delta, x^\dagger) &\le D_{\xi^\dagger}(x^\dagger - \gamma\,\omega, x^\dagger) + \frac{1}{\alpha} D_{Y,p}(y^\delta - A(x^\dagger - \gamma\,\omega), \gamma\,A\,\omega) \\ &\quad + d\left(\varrho_x + \alpha^{\frac{1}{p-1}} R^{\frac{1}{p-1}}\right) \\ &\le G_r \alpha^{\frac{r}{p-1}} R^{\frac{r}{p-1}} + \frac{1}{\alpha} D_{Y,p}(y^\delta - A(x^\dagger - \gamma\,\omega), \gamma\,A\,\omega) \\ &\quad + d\left(\varrho_x + \alpha^{\frac{1}{p-1}} R^{\frac{1}{p-1}}\right). \end{aligned}$$

For $1 < p \le 2$ we have by Corollary 2.2.1 we conclude

$$D_{Y,p}(y^\delta - A(x^\dagger - \gamma\,\omega), \gamma\,A\,\omega) \le \frac{G_p}{p} \|y^\delta - A\,x^\dagger\|^p \le \frac{G_p}{p} \delta^p.$$

Hence, (3.37) holds with $\tilde{K}_1 = \frac{G_p}{p}$. For $p > 2$ we apply Lemma 3.3.2 with $a := 2$, $y_0 := \gamma\,A\,\omega$ and $\Delta y := y^\delta - A\,x^\dagger$. By the triangle inequality and the condition on δ we derive

$$\begin{aligned} \max\{\|t(y^\delta - A\,x^\dagger) + \gamma\,A\,\omega\|, \|\gamma\,A\,\omega\|\} &\le \max\{t\,\delta + \gamma\|A\|\|\omega\|, \gamma\|A\|\|\omega\|\} \\ &\le \delta + \gamma\|A\|\|\omega\| \\ &\le (\tilde{c} + \|A\|)\gamma\|\omega\|. \end{aligned}$$

With $K_1 := (\tilde{c} + \|A\|)^{p-2} C_1/2$ this proves the second part, where C_1 is the constant in the estimate of Lemma 3.3.2. ■

The application of the estimate for $D_{Y,p}(y^\delta - A(x^\dagger - \gamma\omega), \gamma\,A\,\omega)$ is the more sensitive one. In particular, for $p > 2$ and we have to guarantee, that the additional smallness condition on δ of Lemma 3.3.8 is not violated. For fixed $\delta > 0$ this condition is clearly equivalent to $R > 0$. For deriving convergence rates we have to ensure, that this condition holds uniformly, i.e. for $\delta \to 0$, choices $\alpha = \alpha(\delta)$ and $R = R(\delta)$ we have to guarantee $\delta \le \tilde{c}\alpha^{\frac{1}{p-1}} R^{\frac{1}{p-1}}$ for fixed $\tilde{c} > 0$ and all $\delta > 0$ sufficiently small.

Example 3.3.6 *For $p = 2$ the estimate (3.37) of Lemma 3.3.8 reads as*

$$D_{\xi^\dagger}(x_\alpha^\delta, x^\dagger) \le \tilde{K}_1 \frac{\delta^2}{\alpha} + G_r \alpha^r R^r + d\,(\varrho_x + \alpha\,R).$$

We now present convergence rates results proposing an appropriate a-priori parameter choice of the regularization parameter $\alpha = \alpha(\delta)$, see [37, Theorem 4.1].

Theorem 3.3.3 *Assume (A1)-(A6), (A8),(A9) and* $\xi^\dagger \in \overline{\mathcal{M}} \setminus \mathcal{M}$ *has distance function* $d_p(R) := d_p(R;\xi^\dagger)$. *We set* $\tilde{\Theta}_p(R) := d_p(R)^{\frac{p-1}{r}} R^{-1}$.

(i) For $1 < p \leq 2$ *we define the functions*

$$\tilde{\Psi}_p(\alpha) := \left(\alpha\, d_p\left(\tilde{\Theta}_p^{-1}(\alpha)\right)\right)^{\frac{1}{p}} \quad \text{and} \quad \tilde{\Phi}_p(R) := d_p(R)^{\frac{r+p-1}{rp}} R^{-\frac{1}{p}}.$$

(ii) For $p > 0$ *we introduce*

$$\tilde{\Psi}_p(\alpha) := \alpha^{\frac{1}{2}} d_p\left(\tilde{\Theta}_p^{-1}(\alpha)\right)^{\frac{r+2-p}{2r}} \quad \text{and} \quad \tilde{\Phi}_p(R) := d_p(R)^{\frac{r+1}{2r}} R^{-\frac{1}{2}}.$$

Then, the a-priori choice $\alpha := \tilde{\Psi}_p^{-1}(\delta)$ *yields the convergence rate*

$$D_{\xi^\dagger}(x_\alpha^\delta, x^\dagger) = \mathcal{O}\left(d_p\left(\tilde{\Phi}_p^{-1}(\delta)\right)\right) \quad \text{as} \quad \delta \to 0. \tag{3.39}$$

PROOF. We apply Lemma 3.3.8 with $d = d_p(R)$. First we observe that we have to consider only the first three terms of the estimates (3.37) and (3.38), since the term $d_p(R)\alpha^{\frac{1}{p-1}} R^{\frac{1}{p-1}}$ decays faster to zero than $d_p(R)$. By balancing,

$$\alpha^{\frac{r}{p-1}} R^{\frac{r}{p-1}} = d_p(R) \Leftrightarrow \alpha = \frac{d_p(R)^{\frac{p-1}{r}}}{R} = \tilde{\Theta}_p(R) \Leftrightarrow R = \tilde{\Theta}_p^{-1}(\alpha).$$

From now we have to distinguish between $p \leq 2$ and $p > 2$. For $1 < p \leq 2$ we continue with

$$\delta^p \alpha^{-1} = d_p(R) \Leftrightarrow \delta = \left(\alpha\, d_p\left(\tilde{\Theta}_p^{-1}(\alpha)\right)\right)^{\frac{1}{p}} = \tilde{\Psi}_p(\alpha).$$

Hence $\alpha = \tilde{\Psi}_p^{-1}(\delta)$ is the optimal parameter choice. Finally

$$d_p(R) = \delta^p \alpha^{-1} = \delta^p \frac{R}{d_p(R)^{\frac{p-1}{r}}} \Leftrightarrow \delta = \frac{d_p(R)^{\frac{r+p-1}{rp}}}{R^{\frac{1}{p}}} = \tilde{\Phi}_p(R)$$

which provides the convergence rate (3.39).

For $p > 2$ we first assume that the condition on δ of Lemma 3.3.8 is satisfied. Moreover, we have

$$\begin{aligned} d_p(R) = \delta^2 \alpha^{\frac{1}{1-p}} R^{\frac{p-2}{p-1}} &= \delta^2 \alpha^{\frac{1}{1-p}} \left(\frac{d_p(R)^{\frac{1}{r}}}{\alpha^{\frac{1}{p-1}}}\right)^{p-2} \\ &= \delta^2 d_p(R)^{\frac{p-2}{r}} \alpha^{\frac{1}{1-p} - \frac{p-2}{p-1}} \\ &= \delta^2 d_p(R)^{\frac{p-2}{2}} \alpha^{-1} \end{aligned}$$

which gives

$$\delta = d_p(R)^{\frac{r+2-p}{2r}} \alpha^{\frac{1}{2}} = \tilde{\Psi}_p(\alpha)$$

and the parameter choice $\alpha = \tilde{\Psi}_p^{-1}(\delta)$. Then,

$$d_p(R) = \delta^2 \alpha^{\frac{1}{1-p}} R^{\frac{p-2}{p-1}} = \delta^2 R^{\frac{p-2}{p-1}} \frac{R^{\frac{1}{p-1}}}{d_p(R)^{\frac{1}{r}}} = \delta^2 d_p(R)^{-\frac{1}{r}} R,$$

which implies $\delta = R^{-\frac{1}{2}} d_p(R)^{\frac{r+1}{2r}}$ and the corresponding convergence rate (3.39). For the validity of Lemma 3.3.8 it is sufficient to show

$$\frac{\delta}{\alpha^{\frac{1}{p-1}} R^{\frac{1}{p-1}}} = \frac{d_p(R)^{\frac{r+1}{2r}}}{R^{\frac{1}{r}} d_p(R)^{\frac{1}{r}}} = d_p(R)^{\frac{r-1}{2r}} R^{-\frac{1}{2}} \to 0$$

for $\delta \to 0$. Moreover, we observe that

$$d_p(R)^{\frac{r+2-p}{2r}} \alpha^{\frac{1}{2}} = d_p(R)^{\frac{r+2-p}{2r}} \frac{d_p(R)^{\frac{p-1}{2r}}}{R^{\frac{1}{2}}} = \frac{d_p(R)^{\frac{r+1}{2r}}}{R^{\frac{1}{2}}}.$$

Hence, the function $\tilde{\Psi}_p(R)$ is strictly decreasing and consequently, $\tilde{\Psi}_p^{-1}(\alpha)$ is well-defined. ■

Remark 3.3.3 *Similar convergence rates also hold in non-reflexive spaces. By definition of the distance functions we cannot guarantee that $\xi^\dagger \in \mathcal{M}_H(R, d_p(R))$ since the infimum might not be attained. However, $\xi^\dagger \in \mathcal{M}_H(R, d_p(R) + \varepsilon)$ holds for all $\varepsilon > 0$. Assume $1 < p \leq 2$. Repeating the arguments of Corollary 3.3.8 we can prove the estimate*

$$D_{\xi^\dagger}(x_\alpha^\delta, x^\dagger) \leq \tilde{K}_1 \frac{\delta^p}{\alpha} + G_r \alpha^{\frac{r}{p-1}} R^{\frac{r}{p-1}} + (d_p(R) + \varepsilon) \left(\varrho_x + \alpha^{\frac{1}{p-1}} R^{\frac{1}{p-1}} \right)$$

for all $R \geq R_0$ and all $\varepsilon > 0$. Taking the limit $\varepsilon \to 0$ we again can continue as in the proof of Theorem 3.3.3.

The achieved convergence rates essentially depend on the choice of the parameter p only for $p < 2$. This is again devoted the weaker smoothness of the space $\mathcal{Y}$ which was supposed in assumption (A9) for $p < 2$. For $p \geq 2$ the dependence of convergence rate on the choice of p is located only in the specific distance function $d_p(R)$, $R \geq 0$, which – as already mentioned – is rather small. If we again suppose power-type distance functions, then the rate (3.39) does not depend on p anymore. We also observe, that the achieved convergence rates depend on the parameter r which can be considered as measure of the smoothness of the stabilizing functional P.

Example 3.3.7 *We apply the results of Theorem 3.3.3 to some specific situations*

(a) If $r = 2$ the we have $\tilde{\Theta}_p(R) = d_p(R)^{\frac{p-1}{2}} R^{-1}$. Furthermore, for $p \leq 2$ we derive $\tilde{\Phi}_p(R) = d_p(R)^{\frac{p+1}{2p}} R^{-\frac{1}{p}}$. On the other hand, for $p > 2$ we obtain $\tilde{\Psi}_p(R) = \alpha^{\frac{1}{2}} d_p\left(\tilde{\Theta}_p^{-1}(\alpha)\right)^{\frac{4-p}{4}}$ and $\tilde{\Phi}_p(R) = d_p(R)^{\frac{1}{2}} R^{-\frac{1}{2}}$.

(b) For $p = 2$ we get $\tilde{\Theta}_p(R) = d_p(R)^{\frac{1}{r}} R^{-1}$ and $\tilde{\Phi}_p(R) = d_p(R)^{\frac{r+1}{2r}} R^{-\frac{1}{2}}$.

(c) Is, finally, $r = p = 2$ then we derive $\tilde{\Theta}_p(R) = d_p(R)^{\frac{1}{2}} R^{-1}$ and $\tilde{\Phi}_p(R) = d_p(R)^{\frac{3}{4}} R^{-\frac{1}{2}}$.

3.3.5 Improved convergence rates

We now discuss convergence rates under the additional coercivity assumption (A7) for the Bregman distance $D_{\xi^\dagger}(x, x^\dagger)$. Here, we follow the usual strategy: at first we will prove appropriate error bounds which then lead to corresponding convergence rates after a correct balancing of all terms of the error bounds.

First, we present the corresponding error bound result, see [37, Lemma 4.1].

Lemma 3.3.9 *Assume (A1)-(A9) and $\xi^\dagger \in \mathcal{M}_H(R,d)$ for some $R, d \geq 0$.*

- *If $1 < p \leq 2$ then we have the estimate*

$$\|x_\alpha^\delta - x^\dagger\| \leq \tilde{K}_3 \frac{\delta^{\frac{p}{s}}}{\alpha^{\frac{1}{s}}} + K_4 \alpha^{\frac{r}{s(p-1)}} R^{\frac{r}{s(p-1)}} + K_5 d^{\frac{1}{s-1}} + K_6 \alpha^{\frac{1}{p-1}} R^{\frac{1}{p-1}}. \tag{3.40}$$

- *If $p > 2$, and $\delta \leq \tilde{c}\, \alpha^{\frac{1}{p-1}} R^{\frac{1}{p-1}}$ for some constant $\tilde{c} > 0$, then*

$$\|x_\alpha^\delta - x^\dagger\| \leq K_3 \delta^{\frac{2}{s}} \alpha^{\frac{1}{s(1-p)}} R^{\frac{p-2}{s(p-1)}} + K_4 \alpha^{\frac{r}{s(p-1)}} R^{\frac{r}{s(p-1)}} + K_5 d^{\frac{1}{s-1}} + K_6 \alpha^{\frac{1}{p-1}} R^{\frac{1}{p-1}} \tag{3.41}$$

holds. Here, the constants K_i and $\tilde{K}_i$ do not depend on R, d, α and δ.

PROOF. Again we set $\xi^\dagger = A^\star J_p(A\omega) + v$ with $\|\omega\| \leq R^{\frac{1}{p-1}}$ and $\|v\| \leq d$. From the estimate (3.31) and (A7) we conclude

$$\begin{aligned} \|x_\alpha^\delta - x^\dagger\|^s \leq \frac{1}{C_s} D_{\xi^\dagger}(x_\alpha^\delta, x^\dagger) \;&\leq\; \frac{1}{C_s}\left[G_r \alpha^{\frac{r}{p-1}} R^{\frac{r}{p-1}} + \frac{1}{\alpha} D_Y(y^\delta - A(x^\dagger - \gamma\omega), \gamma A\omega) \right] \\ &\quad + \frac{d}{C_s}\left(\|x_\alpha^\delta - x^\dagger\| + \alpha^{\frac{1}{p-1}} R^{\frac{1}{p-1}} \right). \end{aligned}$$

By Young's inequality we estimate

$$\frac{d}{C_s}\left(\|x_\alpha^\delta - x^\dagger\| + \alpha^{\frac{1}{p-1}} R^{\frac{1}{p-1}} \right) \leq 2\frac{s-1}{s}\left(\frac{d}{C_s}\right)^{\frac{s}{s-1}} + \frac{1}{s}\|x_\alpha^\delta - x^\dagger\|^s + \frac{1}{s}\alpha^{\frac{s}{p-1}} R^{\frac{s}{p-1}}.$$

Now we conclude

$$\begin{aligned} \|x_\alpha^\delta - x^\dagger\|^s \;&\leq\; \frac{sG_r}{(s-1)C_s}\alpha^{\frac{r}{p-1}} R^{\frac{r}{p-1}} + \frac{s}{(s-1)C_s}\frac{D_{Y,p}(y^\delta - A(x^\dagger - \gamma\omega), \gamma A\omega)}{\alpha} \\ &\quad + \frac{2}{C_s^{\frac{s}{s-1}}} d^{\frac{s}{s-1}} + \frac{1}{s-1}\alpha^{\frac{s}{p-1}} R^{\frac{s}{p-1}}. \end{aligned}$$

Then the assertions follows directly from Lemma 3.3.8 and the inequality $(a+b)^c \leq a^c + b^c$ for $a, b \geq 0$ and $0 < c < 1$. ■

Example 3.3.8 *We shortly discuss the error bound on basis of some specific situations.*

(a) If $p = 2$ then

$$\|x_\alpha^\delta - x^\dagger\| \leq \tilde{K}_3 \frac{\delta^{\frac{2}{s}}}{\alpha^{\frac{1}{s}}} + K_4 \alpha^{\frac{r}{s}} R^{\frac{r}{s}} + K_5 d_2(R)^{\frac{1}{s-1}} + K_6 \alpha\, R.$$

(b) *If* $r = s = 2$ *then*

$$\|x_\alpha^\delta - x^\dagger\| \le \tilde{K}_3 \frac{\delta^{\frac{p}{2}}}{\alpha^{\frac{1}{2}}} + (K_4 + K_6)\,\alpha^{\frac{1}{p-1}} R^{\frac{1}{p-1}} + K_5 d_p(R)$$

for $p \le 2$. *For* $p > 2$ *we observe*

$$\|x_\alpha^\delta - x^\dagger\| \le K_3 \frac{\delta}{\alpha^{\frac{1}{2(p-1)}}} R^{\frac{p-2}{2(p-1)}} + (K_4 + K_6)\,\alpha^{\frac{1}{p-1}} R^{\frac{1}{p-1}} + K_5 d_p(R).$$

(c) *If* $r = s = p = 2$ *then*

$$\|x_\alpha^\delta - x^\dagger\| \le \tilde{K}_3 \frac{\delta}{\alpha^{\frac{1}{2}}} + (K_4 + K_6)\,\alpha\, R + K_5 d_2(R).$$

We now present improved convergence rates. Again, we have to distinguish between the cases $1 < p \le 2$ and $p > 2$. The accordant results are presented below, see [37, Theorem 5.1].

Theorem 3.3.4 *Assume (A1)-(A9) and* $\xi^\dagger \in \overline{\mathcal{M}} \setminus \mathcal{M}$ *has distance function* $d_p(R) := d_p(R; \xi^\dagger)$. *We set* $\Theta_p(R) := d_p(R)^{\frac{s(p-1)}{r(s-1)}} R^{-1}$.

- *For* $1 < p \le 2$ *we define the functions*

$$\Psi_p(\alpha) := \left(\alpha\, d_p\left(\Theta_p^{-1}(\alpha)\right)^{\frac{s}{s-1}}\right)^{\frac{1}{p}} \quad \text{and} \quad \Phi_p(R) := d_p(R)^{\frac{s(r+p-1)}{(s-1)rp}} R^{-\frac{1}{p}}.$$

(ii) *For* $p > 2$ *we set*

$$\Psi_p(\alpha) := \alpha^{\frac{1}{2}} d_p\left(\Theta_p^{-1}(\alpha)\right)^{\frac{s(r+2-p)}{2r(s-1)}} \quad \text{and} \quad \Phi_p(R) := d_p(R)^{\frac{s(r+1)}{2r(s-1)}} R^{-\frac{1}{2}}.$$

Then, the a-priori choice $\alpha := \Psi_p^{-1}(\delta)$ *yields the convergence rate*

$$\|x_\alpha^\delta - x^\dagger\| = \mathcal{O}\left(d_p\left(\Phi_p^{-1}(\delta)\right)^{\frac{1}{s-1}}\right) \quad \text{as } \delta \to 0. \tag{3.42}$$

Proof. First we observe that the term $\alpha^{\frac{1}{p-1}} R^{\frac{1}{p-1}}$ decays to zero at least as fast as $\alpha^{\frac{r}{s(p-1)}} R^{\frac{r}{s(p-1)}}$ since we always have $r \le s$. Comparing the estimates of Lemma 3.3.8 and Lemma 3.3.9 we observe that balancing the terms of the error estimates (3.38) and (3.40) now are similar to the proof of Theorem 3.3.3. We only have to replace the distance functions $d_p(R)$ by $d_p(R)^{\frac{s}{s-1}}$. Finally, it is a short calculation that for $p > 2$ the assumption of Lemma 3.3.8 on the noise-level δ is fulfilled. ■

Example 3.3.9 *We apply the convergence rate results of Theorem 3.3.4.*

(a) *For* $p = 2$ *we observe* $\Theta_2(R) = d_p(R)^{\frac{s}{r(s-1)}} R^{-1}$.

(b) *For* $r = s = 2$ *we see that* $\Theta_p(R) = d_p(R)^{p-1} R^{-1}$. *If* $p \le 2$ *we calculate* $\Theta_p(R) = d_p(R)^{\frac{p+1}{p}} R^{-\frac{1}{p}}$. *On the other hand, for* $p > 2$ *we have* $\Psi_p(\alpha) = \alpha^{\frac{1}{2}} d_p\left(\Theta_p^{-1}(\alpha)\right)^{\frac{4-p}{2}}$ *and* $\Phi_p(R) = d_p(R)^{\frac{3}{2}} R^{-\frac{1}{2}}$.

(c) Finally, for $p = r = s = 2$ we obtain $\Theta_2(R) = d_2(R)R^{-1}$.

Clearly, the convergence rates results of Theorem 3.3.3 and 3.3.4 need some interpretation. For given $\xi \in \mathcal{X}^*$ and distance function $d_p(R;\xi)$, $R \geq 0$, we cannot state the corresponding convergence rates explicitly in general. The reason for it is devoted the observation that $\tilde{\Phi}_p^{-1}(\delta)$ and $\Phi_p^{-1}(\delta)$ are not explictly calculable for arbitrary distance functions. However, this will change when we suppose power-type distance functions. We deal with the following example.

Example 3.3.10 *Let us suppose $\mathcal{X}$ and $\mathcal{Y}$ to be Hilbert spaces and $P(x) := \frac{1}{2}\|x\|^2$. We recall that this implies $\xi = x^\dagger$, $D_{\xi^\dagger}(x, x^\dagger) = \frac{1}{2}\|x - x^\dagger\|^2$ for $x \in \mathcal{X}$ and $r = s = 2$. Additionally we choose $p = 2$. We furthermore assume now that the distance function $d_p(R;\xi)$ has the form*

$$d_2(R; x^\dagger) := C\, R^{\frac{\nu}{\nu-1}}, \qquad R \geq R_0 > 0,$$

for some constant $C > 0$ and parameter $0 < \nu < 1$. We already motivated the specific structure of the exponent: if $x^\dagger \in \mathcal{R}((A^*A)^\nu)$ then $d_2(R; x^\dagger)$ is of above structure. Then we have for the function $\tilde{\Theta}_2(R)$ in Theorem 3.3.3 (with some generic constant $\mathcal{C} > 0$)

$$\tilde{\Theta}_2(R) := d_2(R)^{\frac{1}{2}} R^{-1} = \mathcal{C}\, R^{\frac{\nu}{2(\nu-1)}-1} = \mathcal{C}\, R^{\frac{2-\nu}{2(\nu-1)}} = \alpha,$$

which implies $R = \mathcal{C}\alpha^{\frac{2(\nu-1)}{2-\nu}} = \tilde{\Theta}_2^{-1}(\alpha)$. Hence

$$\tilde{\Psi}_2(\alpha) = \sqrt{\alpha\, d_2\left(\tilde{\Theta}_2^{-1}(\alpha)\right)} = \mathcal{C}\alpha^{\frac{1}{2}\left(1+\frac{\nu}{\nu-1}\frac{2(\nu-1)}{2-\nu}\right)} = \mathcal{C}\alpha^{\frac{2+\nu}{2(2-\nu)}},$$

which provides a parameter choice $\alpha \sim \delta^{\frac{2(2-\nu)}{2+\nu}}$. Moreover

$$\tilde{\Phi}_2(R) = d_2(R)^{\frac{3}{4}} R^{-\frac{1}{2}} = \mathcal{C}\, R^{\frac{3\nu}{4(\nu-1)}-\frac{1}{2}} = \mathcal{C}\, R^{\frac{\nu+2}{4(\nu-1)}}$$

and

$$d_2\left(\tilde{\Phi}_2^{-1}(\delta)\right) = \mathcal{C}\,\delta^{\frac{\nu}{\nu-1}\frac{4(\nu-1)}{\nu+2}} = \mathcal{C}\,\delta^{\frac{4\nu}{\nu+2}}$$

hold. This yields a convergence rate $\|x_\alpha^\delta - x^\dagger\| \sim \delta^{\frac{2\nu}{\nu+2}}$, which is not the optimal one. On the other hand, in Theorem 3.3.4 we introduced

$$\Theta_2(R) = d_2(R)R^{-1} = \mathcal{C}\, R^{\frac{\nu}{\nu-1}-1} = \mathcal{C}\, R^{\frac{1}{\nu-1}} = \alpha$$

or $R = \mathcal{C}\,\alpha^{\nu-1}$. Hence

$$\Psi_2(\alpha) = \alpha^{\frac{1}{2}} d_2\left(\Theta_2^{-1}(\alpha)\right) = \mathcal{C}\,\alpha^{\frac{1}{2}+\frac{\nu}{\nu-1}(\nu-1)} = \mathcal{C}\,\alpha^{\frac{2\nu+1}{2}},$$

which implies the well-known optimal parameter choice $\alpha = \mathcal{C}\,\delta^{\frac{2}{2\nu+1}}$. Finally,

$$\Phi_2(R) = d_2(R)^{\frac{3}{2}} R^{-\frac{1}{2}} = \mathcal{C}\, R^{\frac{3\nu}{2(\nu-1)}-\frac{1}{2}} = \mathcal{C}\, R^{\frac{2\nu+1}{2(\nu-1)}} = \delta,$$

which provides

$$R = \mathcal{C}\,\delta^{\frac{2(\nu-1)}{2\nu+1}} = \Phi_2^{-1}(\delta).$$

Then we derive the convergence rate

$$\|x_\alpha^\delta - x^\dagger\| \sim d_2\left(\Phi_2^{-1}(\delta)\right) = \mathcal{C}\,\delta^{\frac{2\nu}{2\nu+1}} = \mathcal{C}\,\delta^{\frac{2\nu}{2\nu+1}}. \tag{3.43}$$

The rate (3.43) is known to be the optimal one for any regularization methods if $x^\dagger \in \mathcal{R}((A^*A)^\nu)$ for $0 < \nu \le 1$. Hence, power-type source conditions and distance functions provide the same (optimal) convergence rates.

More generally, for arbitrary $s \ge r > 1$ we achieve for $1 < p \le 2$ the function

$$\Phi_p(R) = \mathcal{C}\,R^{\frac{\nu s(r+p-1)}{(\nu-1)(s-1)rp}-\frac{1}{p}} = \mathcal{C}\,R^{\frac{s\nu(p-1)+r(s+\nu-1)}{(\nu-1)(s-1)rp}}$$

and consequently the convergence rate

$$\|x_\alpha^\delta - x^\dagger\| \sim \delta^{\frac{rp\nu}{s\nu(p-1)+r(s+\nu-1)}}.$$

In particular, for $r = s = 2$ this reduces to $\|x_\alpha^\delta - x^\dagger\| \sim \delta^{\frac{p\nu}{p\nu+1}}$ which - in comparison to (3.43) - shows the influence of the lower smoothness of the space $\mathcal{Y}$ on the convergence rate. For $p > 2$ we have

$$\Phi_p(R) = \mathcal{C}\,R^{\frac{\nu s(r+1)}{2r(\nu-1)(s+1)}-\frac{1}{2}} = \mathcal{C}\,R^{\frac{\nu(s+r)+r(s-1)}{2r(s-1)(\nu-1)}}$$

and the corresponding convergence rate

$$\|x_\alpha^\delta - x^\dagger\| \sim \delta^{\frac{2\nu r}{\nu(s+r)+r(s-1)}}.$$

3.3.6 On an a-posteriori parameter choice

We now discuss strategies for choosing the regularization parameter α without knowledge of the (approximate) source condition in a proper way. Known regularization theory in Hilbert space we cannot expect to obtain better convergence rates than $D_{\xi^\dagger}(x_\alpha^\delta, x^\dagger) = \mathcal{O}(\delta)$ as $\delta \to 0$ by using the discrepancy principle. Therefore it makes no sense to discuss this strategy here. On the other hand for linear operators in Hilbert spaces some modified discrepancy principles are known leading to optimal convergence rates, see [20].

Here, we focus on the applicability of the balancing principle as a-posteriori parameter choice for deriving the convergence rates (3.34) and (3.42) of Theorem 3.3.1 and 3.3.4 respectively. We recall, the the coercivity condition (A7) is necessarry for the application of the balancing principle.

We start with the case $1 < p \le 2$. Assume $\xi^\dagger \in \mathcal{X}^*$ satisfies the source condition (3.29) for some $\omega \in \mathcal{X}$ with $\|\omega\| \le R^{\frac{1}{p-1}}$. Then, from the estimate (3.37) we derive with $d = 0$

$$C_s\|x_\alpha^\delta - x^\dagger\|^s \le D_{\xi^\dagger}(x_\alpha^\delta, x^\dagger) \le \tilde{K}_2\frac{\delta^p}{\alpha} + G_r\alpha^{\frac{2}{p-1}}R^{\frac{2}{p-1}}$$

or respectively

$$\|x_\alpha^\delta - x^\dagger\| \le \left(\frac{\tilde{K}_2}{C_s}\right)^{\frac{1}{s}} \frac{\delta^{\frac{p}{s}}}{\alpha^{\frac{1}{s}}} + \left(\frac{G_r}{C_s}\right)^{\frac{1}{s}} R^{\frac{1}{p-1}} \alpha^{\frac{1}{p-1}}.$$

On the other hand, for $\xi^\dagger \in \overline{\mathcal{M}} \setminus \mathcal{M}$ and distance function $d_p(R) = d_p(R; \xi^\dagger)$ we derive from (3.37)

$$\begin{aligned} \|x_\alpha^\delta - x^\dagger\| &\le \left(\frac{s\tilde{K}_2}{(s-1)C_s}\right)^{\frac{1}{s}} \frac{\delta^{\frac{p}{s}}}{\alpha^{\frac{1}{s}}} + K_3 \alpha^{\frac{r}{s(p-1)}} R^{\frac{r}{s(p-1)}} + K_5 d_p(R)^{\frac{1}{s-1}} + K_6 \alpha^{\frac{1}{p-1}} R^{\frac{1}{p-1}} \\ &= \left(\frac{s\tilde{K}_2}{(s-1)C_s}\right)^{\frac{1}{s}} \frac{\delta^{\frac{p}{s}}}{\alpha^{\frac{1}{s}}} + \left(K_3 + K_5 + K_6 d_p\left(\Theta_p^{-1}(\alpha)\right)^{\frac{s-r}{r(s-1)}}\right) d_p\left(\Theta_p^{-1}(\alpha)\right)^{\frac{1}{s-1}} \end{aligned}$$

after the first balancing step in the proof of Theorem 3.3.3. We define the function

$$\psi(\alpha) := \left(\frac{s\tilde{K}_2}{(s-1)C_s}\right)^{\frac{1}{s}} \frac{\delta^{\frac{p}{s}}}{\alpha^{\frac{1}{s}}}, \qquad \alpha > 0.$$

Hence, in both cases, we can divide the error $\|x_\alpha^\delta - x^\dagger\|$ for given $\delta > 0$ into

$$\|x_\alpha^\delta - x^\dagger\| \le \psi(\alpha) + \phi(\alpha)$$

with known decreasing function $\psi(\alpha)$ depending also on the noise-level δ and an unknown non-decreasing (index) function $\phi(\alpha)$ which depends on the (approximative) source condition but not on the noise-level δ. This is again exactly the situation which allows us to apply the balancing principle. For the presentation of our convergence rates analysis based on a-priori parameter choice in the previous sections we had to distinguish between satisfied and violated source condition (3.29). For the applicability of the balancing principle it turns out that we do not need to know which situation actually occurs. The reason for that can be found in the fact that the known function $\psi(\alpha)$ of the error estimate is in both cases (up to a constant) the same. The (possible) violation of the source condition (3.29) only reflects the unknown function $\phi(\alpha)$ which knowledge is not needed for the balancing principle.

For the application of the balancing principle for $p > 2$ we have to introduce an additional consideration. In particular, the term $K_3 \delta^{\frac{2}{s}} \alpha^{\frac{1}{s(1-p)}} R^{\frac{p-2}{s(p-1)}}$ of the error estimate (3.38) depends also on the unknown (approximative) source condition. Therefore we apply Young's inequality with some $\mu > 1$ to obtain

$$K_3 \delta^{\frac{2}{s}} \alpha^{\frac{1}{s(1-p)}} R^{\frac{p-2}{s(p-1)}} = K_3 \frac{\delta^{\frac{2}{s}}}{\alpha^{\frac{1}{s}}} (\alpha R)^{\frac{p-2}{s(p-1)}} \le \frac{\mu-1}{\mu} \left(\frac{\delta^2}{\alpha}\right)^{\frac{\mu}{s(\mu-1)}} + \frac{1}{\mu} K_3^\mu (\alpha R)^{\frac{\mu(p-2)}{s(p-1)}}.$$

The first term does not depend on any (approximate) source condition. For $p - 2 < r$ we can choose $\mu = \frac{r}{p-2}$. Then the second term is of the same order than the second term in the estimate (3.37). The error estimate (3.38) with $d = 0$ for satisfied source condition (3.29) we can treat similarly. So we again apply the balancing principle with

$$\psi(\alpha) := \frac{\mu - 1}{\mu} \left(\frac{\delta^2}{\alpha}\right)^{\frac{\mu}{s(\mu-1)}} = \frac{r+2-p}{r} \left(\frac{\delta^2}{\alpha}\right)^{\frac{r}{s(r+2-p)}}, \qquad \alpha > 0.$$

It turns out here, that the application of Young's inequality does not destroy the convergence rates. In particular, using the balancing principle for the choice of the regularization parameter α leads again to the convergence rates (3.34) or (3.39) respectively, depending whether source condition (3.29) holds for some $R > 0$ or only an approximate source condition (3.30) is fulfilled. For $p - 2 \geq r$ we have $\mu > \frac{r}{p-2}$. Hence, the exponent in the second term is larger than $\frac{r}{s(p-1)}$ which will lead to lower lower convergence rates. So it is up to now an open problem to find an a-posteriori parameter choice yielding the optimal convergence rates (3.34) or (3.39) respectively for that situation.

3.4 Minimization of Tikhonov functionals

In the following we discuss algorithms for finding a minimizer x_α^δ of the corresponding Tikhonov functional T_{α,y^δ}. We restrict our considerations to the minimization problem

$$T_{\alpha,y^\delta}(x) := \frac{1}{p}\|A\,x - y^\delta\|^p + \frac{\alpha}{s}\|x\|^s \to \min \quad \text{subject to} \quad x \in \mathcal{X}. \tag{3.44}$$

In particular we consider only linear operators $A : \mathcal{X} \longrightarrow \mathcal{Y}$ and the penality functional $P(x) := s^{-1}\|x\|^s$. Since the functional T_{α,y^δ} is strictly convex there exists a unique minimizer $x_\alpha^\delta \in \mathcal{X}$ of this problem.

Throughout this section we assume the following:

- The space $\mathcal{X}$ is smooth and s-convex for some $s \in [2, \infty)$. Hence the dual space $\mathcal{X}^*$ is s^*-smooth. In particular, we connect the choice of the exponent of the penalty term to the convexity of the space $\mathcal{X}$. Moreover, let $J_s : \mathcal{X} \longrightarrow \mathcal{X}^*$ and $J_{s^*}^* : \mathcal{X}^* \longrightarrow \mathcal{X}$ denote the corresponding duality mappings (with gauge functions $t \mapsto t^{s-1}$ and $t \mapsto t^{s^*-1}$ respectively). We remember, that by the properties of duality mappings and the space $\mathcal{X}$ we have $J_{s^*}^* = J_s^{-1}$.
- The space $\mathcal{Y}$ is assumed to be smooth, i.e. the duality mapping $J_p : \mathcal{Y} \longrightarrow \mathcal{Y}^*$ is always single valued for some $p > 1$.

In the following we we deal with both gradient-based methods and approaches of Newton-type.

3.4.1 Gradient-based methods

Gradient methods in Banach spaces need some special treatment. Simple calculations show that the gradient $T'_{\alpha,y^\delta}(x)$ at $x \in \mathcal{X}$ is given as

$$T'_{\alpha,y^\delta}(x) := A^\star J_p(A\,x - y^\delta) + \alpha J_s(x) \in \mathcal{X}^*.$$

As opposite to gradient methods in Hilbert spaces we have to take into account that the gradient belongs to the dual space $\mathcal{X}^*$ of $\mathcal{X}$. Therefore, the gradients have to be transferred into the original space $\mathcal{X}$ for calculating the new iterate. Here, duality mappings are the natural choice for this transportation from $\mathcal{X}$ into $\mathcal{X}^*$ and vice versa. Moreover, we now we have two possibilities for defining the update. On the one hand, for given initial guess $x_0 \in \mathcal{X}$ we can iterate

$$x_{n+1} := x_n - \mu_n J^*_{s^*}\left(T'_{\alpha,y^\delta}(x_n)\right), \quad n = 0, 1, \ldots.$$

with proper chosen step size μ_n. Alternatively, we can perform choice of the step size μ_n and the update in the dual space $\mathcal{X}^*$ and return afterwards to the solution space $\mathcal{X}$. The general scheme is given as follows: for given $x_0^* \in \mathcal{X}^*$ and $x_0 := J^*_{s^*}(x_0^*)$ we iterate

$$\begin{aligned} x^*_{n+1} &:= x^*_n - \mu_n T'_{\alpha,y^\delta}(x_n), \\ x_{n+1} &:= J^*_{s^*}(x^*_{n+1}), \qquad n = 0, 1, \ldots. \end{aligned}$$

The choice of the parameter μ_n is essential for the speed of convergence of the algorithm. Choosing $\mu_n \equiv$ *const.* the convergence may become rather slow. In arbitrary Banach spaces we even cannot guarantee linear convergence $x_n \to x^\delta_\alpha$ as $n \to \infty$.

We present some results.

a) Iterating in the dual space

For $\tilde{x}, x \in X$ we define the Bregman distance

$$\begin{aligned} \Delta_s(\tilde{x}, x) &:= \frac{1}{s}\|\tilde{x}\|^s - \frac{1}{s}\|x\|^s - \langle J_s(x), \tilde{x} - x\rangle \\ &= \frac{1}{s}\|\tilde{x}\|^s + \frac{1}{s^*}\|J_s(x)\|^{s^*} - \langle J_s(x), \tilde{x}\rangle, \end{aligned}$$

for the functional $x \mapsto \frac{1}{s}\|x\|^s$. Following the idea in [56] we derive a proper choice of the step size parameter μ_n in the algorithm. To shorten the notation we define $\Delta_n := \Delta_s(x^\delta_\alpha, x_n)$ and $\Psi^*_n := T'_{\alpha,y^\delta}(x_n)$. Then with $\mu = \mu_n$ we have

$$\begin{aligned} \Delta_{n+1} &= \frac{1}{s}\|x^\delta_\alpha\|^s + \frac{1}{s^*}\|x^*_{n+1}\|^{s^*} - \langle x^*_n - \mu\,\Psi^*_n, x^\delta_\alpha\rangle \\ &\le \frac{1}{s}\|x^\delta_\alpha\|^s - \langle x^*_n, x^\delta_\alpha\rangle + \mu\langle \Psi^*_n, x^\delta_\alpha\rangle + \frac{1}{s^*}\|x^*_n\|^{s^*} - \mu\langle x_n, \Psi^*_n\rangle + \frac{G^*_{s^*}}{s^*}\|\Psi^*_n\|^{s^*}\mu^{s^*} \\ &= \Delta_n - \mu\langle \Psi^*_n, x_n - x^\delta_\alpha\rangle + \frac{G^*_{s^*}}{s^*}\|\Psi^*_n\|^{s^*}\mu^{s^*}. \end{aligned}$$

The constant $G^*_{s^*}$ comes from the s^*-smoothness of the dual space $\mathcal{X}^*$. Moreover, with $T'_{\alpha,y^\delta}(x^\delta_\alpha) = 0$ we derive

$$\begin{aligned} \langle \Psi^*_n, x^\delta_\alpha - x_n\rangle &= \langle \Psi^*_n - T'_{\alpha,y^\delta}(x^\delta_\alpha), x^\delta_\alpha - x_n\rangle \\ &= \underbrace{\big\langle J_p(A\,x_n - y^\delta) - J_p(A\,x^\delta_\alpha - y^\delta), A\,x^\delta_\alpha - A\,x_n\big\rangle}_{\le 0} \\ &\quad + \alpha\langle x^*_n - J_s(x^\delta_\alpha), x^\delta_\alpha - x_n\rangle \\ &\le -\alpha\langle x^*_n - J_s(x^\delta_\alpha), x_n - x^\delta_\alpha\rangle \\ &= -\alpha\left(\Delta_n + \Delta_s(x_n, x^\delta_\alpha)\right) \le -\alpha\,\Delta_n \end{aligned}$$

by using the monotonicity of duality mappings and the non-negativity of the Bregman distance $\Delta_s(x_n, x_\alpha^\delta)$. Hence, we end at

$$\Delta_{n+1} \leq (1 - \alpha\,\mu)\,\Delta_n + \frac{G_{s^*}^*}{s^*}\|\Psi_n^*\|^{s^*}\mu^{s^*}.$$

Of course, we do not know Δ_n exactly. We therefore assume that we know at least an upper bound $D_n > 0$ satisfying $\Delta_n \leq D_n$. Then we can rewrite the above inequality as

$$\Delta_{n+1} \leq \max\{1 - \alpha\,\mu, 0\}\,D_n + \frac{G_{s^*}^*}{s^*}\|\Psi_n^*\|^{s^*}\mu^{s^*}.$$

The right hand side becomes minimal if we choose $\mu = \mu_n$ as

$$\mu_n := \min\left\{\left(\frac{\alpha}{G_{s^*}^*}\frac{D_n}{\|\Psi_n^*\|}\right)^{\frac{1}{s^*-1}}, \frac{1}{\alpha}\right\}.$$

We therefore introduce the following algorithm:

Algorithm 3.4.1

(S0) Choose $x_0 \in \mathcal{X}$, $x_0^* := J_s(x_0)$, $D_0 > 0$ *such that* $\Delta_s(x_\alpha^\delta, x_0) \leq D_0$, $n := 0$

(S1) Calculate $\Psi_n^* := T'_{\alpha,y^\delta}(x_n)$, *STOP if* $\Psi_n^* = 0$ *else set*

$$\begin{aligned} \mu_n &:= \min\left\{\left(\frac{\alpha}{G_{s^*}^*}\frac{D_n}{\|\Psi_n^*\|}\right)^{\frac{1}{s^*-1}}, \frac{1}{\alpha}\right\} \\ D_{n+1} &:= (1 - \mu_n\alpha)D_n + \mu_n^{s^*}\frac{G_{s^*}^*}{s^*}\|\Psi_n^*\|^{s^*} \end{aligned}$$

(S3) Set $x_{n+1}^* := x_n^* - \mu_n\Psi_n^*$, $x_{n+1} := J_{s^*}^*(x_{n+1}^*)$

(S4) Set $n := n + 1$, *go to step (S1).*

Then the following result can be proven, see e.g. [56] where this choice of the step size was derived in a slightly different way.

Proposition 3.4.1 *Let $\mathcal{X}$ be a s-convex Banach space and $\mathcal{Y}$ an arbitrary Banach space. Then the sequence $\{x_n\}$ generated by Algorithm 3.4.1 converges strongly to the minimizer x_α^δ of the Tikhonov functional T_{α,y^δ} at least with the rate*

$$\|x_n - x_\alpha^\delta\| \leq C\,n^{-\frac{1}{s(s-1)}}.$$

If additionally the spaces $\mathcal{X}$ and $\mathcal{Y}$ are 2-convex and $p = s = 2$ then the sequence $\{x_n\}$ converges linearly to x_α^δ.

Sketch of the Proof. The proof essentially bases on the following steps, see [56]:

1. Observe that the sequence $\{D_n\}$ is monotone decreasing. It is clear by induction that $\Delta_n \leq D_n$ holds for all n.

2. Show, that $D_{n+1}^{1-s} - D_n^{1-s} \geq \mathcal{C} D_n^{1-s}$ for some constant $\mathcal{C} > 0$ which does not depend on the iteration number n. This gives $D_n \to 0$ as $n \to \infty$.
3. Apply the s-convexity of the space $\mathcal{X}$ to obtain the desired rate. Under the additional assumption one can show, that the sequence $\{D_n\}$ decays linearly which also implies the linear decay of $\{\|x_n - x_\alpha^\delta\|\}$. ■

We presented this idea here in detail since we will apply a similar argumentation for proving descent properties of iterative regularization methods in the next chapter. We also refer to [9] for an alternative algorithm.

b) The method of steepest descent

A widely used search of the step size parameter is the method of steepest descent (also called Cauchy step size). The idea is simple: for given search direction (using the gradient in Hilbert spaces) we choose the parameter $\mu = \mu_n$ in such a way that the underlying functional becomes minimal within this direction. Hence we have to solve an one-dimensional minimization problem in each iteration step which is often easy to handle. We generalize the method of steepest descent to a Banach space setting.

Algorithm 3.4.2

(S0) Choose $x_0 \in \mathcal{X}$, set $n := 0$.

(S1) Calculate $\Psi_n^* := T'_{\alpha,y^\delta}(x_n)$. If $\Psi_n^* = 0$ then STOP, else do a line search to find $\mu_n > 0$ such that

$$T_{\alpha,y^\delta}\left(x_n - \mu_n J_{s^*}^*(\Psi_n^*)\right) = \min_{\mu \in \mathbb{R}} T_{\alpha,y^\delta}\left(x_n - \mu\, J_{s^*}^*(\Psi_n^*)\right).$$

(S2) Set $x_{n+1} := x_n - \mu_n J_{s^*}^*(\Psi_n^*)$

(S3) Set $n := n+1$ and go to step (S1).

We can present the following convergence result, see [9].

Proposition 3.4.2 *The sequence $\{x_n\}$ generated by Algorithm 3.4.2 converges strongly to the minimizer x_α^δ of the Tikhonov functional T_{α,y^δ}.*

We shortly discuss the line search step (S1). We define the function $f_n := \mathbb{R} \longrightarrow [0,\infty)$ as.

$$f_n(\mu) := T_{\alpha,y^\delta}\left(x_n - \mu\, J_{s^*}^*(\Psi_n^*)\right)$$

This function is strictly convex and differentiable with derivative

$$f'_n(\mu) = -\left\langle T'_{\alpha,y^\delta}\left(x_n - \mu\, J_{s^*}^*(\Psi_n^*)\right), J_{s^*}^*(\Psi_n^*)\right\rangle.$$

Furthermore, we have $f'_n(0) = -\|T'_{\alpha,y^\delta}(x_n)\|^s < 0$. Since f'_n is increasing by the monotonicity of the duality mappings we conclude $\mu_n > 0$. This is evident since with $\Delta x_n := J_{s^*}^*(T'_{\alpha,y^\delta}(x_n))$ we see

$$\begin{aligned} f'_n(\mu) &= -\left\langle A^* J_p\left(A(x_n - \mu\,\Delta x_n) - y^\delta\right) + \alpha J_s\left(x_n - \mu\,\Delta x_n\right), \Delta x_n\right\rangle \\ &= -\left(\left\langle J_p\left(A\,x_n - y^\delta - \mu\, A\,\Delta x_n\right), A\,\Delta x_n\right\rangle + \alpha\left\langle J_s\left(x_n - \mu\,\Delta x_n\right), \Delta x_n\right\rangle\right). \end{aligned}$$

So we can apply [97, Beispiel 24.4] which says that the monotonicity of the duality mapping $J_s : \mathcal{X} \longrightarrow \mathcal{X}^*$ implies that the mapping $t \mapsto \langle J_s(x + t\,\tilde{x}), \tilde{x}\rangle$ is increasing for all $x, \tilde{x} \in \mathcal{X}$. The same holds for $J_p : \mathcal{Y} \longrightarrow \mathcal{Y}^*$. So the line search is easy to implement.

3.4.2 Newton-type methods

We now deal with Newton-type methods. One of the main application of Tikhonov regularization in Banach spaces is the use of non-smooth penalty functionals P. In particular, in sparsity reconstruction the choice $\mathcal{X} = L^q$ with q close to 1 is suggested for a better estimation of non-smooth solutions $x^\dagger \in \mathcal{X}$. So, Newton's method in its original version cannot applied directly. We refer e.g. to [25] for the application of a semi-smooth Newton method in that situation. On the other hand we can give an alternative approaches which uses the fact that in this situation the dual space $\mathcal{X}^* = L^{\frac{q}{q-1}}$ has higher smoothness.

Starting point is the necessary (and sufficient) optimality condition

$$T'_{\alpha,y^\delta}(x) = A^* J_p(A\,x - y^\delta) + \alpha\, J_s(x) = 0 \in \mathcal{X}^*. \tag{3.45}$$

We present two approaches for solving this equation.

a) The classical method

Let $x \in \mathcal{X}$ be given. In order to apply Newton's method in its classical form we need the following assumptions:

(i) The duality mapping J_s is Gâteaux-differentiable in $x \in \mathcal{X}$, i.e. there exists a linear bounded operator $J'_s : \mathcal{X} \longrightarrow \mathcal{X}^*$ such that

$$J'_s(x)\,h = \lim_{\varepsilon\to 0} \frac{J_s(x + \varepsilon\,h) - J_s(x)}{\varepsilon} \qquad \forall\, h \in \mathcal{X}.$$

(ii) There exists a linear bounded operator $J'_p(A\,x - y^\delta) : \mathcal{R}(A) \subseteq \mathcal{Y} \longrightarrow \mathcal{Y}^*$ such that

$$J'_p(A\,x - y^\delta)\,\tilde{h} = \lim_{\varepsilon\to 0} \frac{J_p(A\,x - y^\delta + \varepsilon\,\tilde{h}) - J_p(A\,x - y^\delta)}{\varepsilon} \qquad \forall\, \tilde{h} \in \mathcal{R}(A).$$

In particular, the conditions (i) and (ii) are fulfilled for all $x \in \mathcal{X}$ if $\mathcal{X}$ and $\mathcal{Y}$ are supposed to be 2-smooth spaces. Under these assumptions T'_{α,y^δ} is Gâteaux-differentiable in $x \in \mathcal{X}$ and its derivative $T''_{\alpha,y^\delta}(x) : \mathcal{X} \longrightarrow \mathcal{X}^*$ is given as

$$T''_{\alpha,y^\delta}(x) := A^* J'_p(A\,x - y^\delta)A + \alpha\, J'_s(x).$$

We recall that a operator $B : \mathcal{X} \longrightarrow \mathcal{X}^*$ is called

- positive if $\langle B\,x, x\rangle \geq 0$,
- strictly positive if $\langle B\,x, x\rangle > 0$ and
- strongly positive if $\langle B\,x, x\rangle \geq C\,\|x\|^2$ for some constant $C > 0$ and all $x \in \mathcal{X}$.

Now we can present the following lemma.

Lemma 3.4.1 *Under the condition above the operator $T''_{\alpha,y^\delta}(x)$ is positive For $s=2$ we even have strong positivity.*

PROOF. Let $h \in \mathcal{X}$ with $h \neq 0$ be arbitrary chosen. We have the properties of duality mappings, see [96], that

$$\langle J_s(x+\varepsilon\, h) - J_s(x), h\rangle = \frac{1}{\varepsilon}\langle J_s(x+\varepsilon\, h) - J_s(x), \varepsilon\, h\rangle \geq \frac{K}{\varepsilon}\|\varepsilon\, h\|^s = K\,\varepsilon^{s-1}\|h\|^s,$$

which holds for all $\varepsilon > 0$ and a constant $K > 0$ which does not depend on ε, h and x. Since $s \geq 2$ we conclude from the differentiability assumption

$$\infty > \langle J_s'(x)\, h, h\rangle = \lim_{\varepsilon\to 0}\frac{1}{\varepsilon^2}\left(\langle J_s(x+\varepsilon\, h) - J_s(x), \varepsilon\, h\rangle\right) \geq \begin{cases} K\|h\|^2, & s=2, \\ 0, & s>2. \end{cases}$$

From the monotonicity of duality mapping we further conclude by a similar calculation that

$$\infty > \langle A^* J_p'(A\,x - y^\delta)A\,h, h\rangle = \langle J_p'(A\,x - y^\delta)A\,h, A\,h\rangle \geq 0.$$

This proves the lemma. ■

Taking into account, that $T''_{\alpha,y^\delta}(x)$ might not exists for all $x \in \mathcal{X}$ we present the following algorithm.

Algorithm 3.4.3

(S0) Choose $x_0 \in \mathcal{X}$, set $n := 0$.

(S1) If $T'_{\alpha,y^\delta}(x_n) = 0$ then STOP.

(S2) Choose a strictly positive, self-adjoint and bounded operator $H_n : \mathcal{X} \longrightarrow \mathcal{X}^*$ and determine the search direction $h = h_n \in \mathcal{X}$ as weak solution of the linear equation

$$H_n\, h = -T'_{\alpha,y^\delta}(x_n).$$

(S3) Find a step size $\mu_n > 0$ such that with $x_{n+1} := x_n + \mu_n\, h$ we have $T_{\alpha,y^\delta}(x_{n+1}) < T_{\alpha,y^\delta}(x_n)$.

(S4) Set $n := n+1$ and go to step (S1).

We examine the steps (S2) and (S3). By a version of the Lax-Milgram theorem [63] for symmetric bilinear forms in Banach spaces we have existence and uniqueness of $h_n \in \mathcal{X}$ (depending stable on $T'_{\alpha,y^\delta}(x_n)$). In particular, if

- the second Gâteaux derivative $T''_{\alpha,y^\delta}(x_n) : \mathcal{X} \longrightarrow \mathcal{X}^*$ exists and
- is strictly positive

we can set $H_n := T''_{\alpha,y^\delta}(x_n)$. Otherwise we choose H_n in such a way that H_n might represents an approximation of $T''_{\alpha,y^\delta}(x_n)$ in some sense. Moreover, the strict positivity of the operator H_n guarantees that h_n is a descent direction of T_{α,y^δ} in x_n. In particular, the steps (S2) and (S3) in the algorithm are well-defined. We refer also to the wide literature on Newton methods for discussing the convergence and speed of convergence of this algorithm, see e.g. [15] an the references therein.

b) Newton's method in the dual space

The idea is based on the following situation: we set $x^* := J_s(x) \Leftrightarrow x = J^*_{s^*}(x^*)$ and rewrite equation (3.45) as

$$T'_{\alpha,y^\delta}\left(J^*_{s^*}(x^*)\right) = A^* J_p\left(A\, J^*_{s^*}(x^*) - y^\delta\right) + \alpha\, x^* = 0.$$

If we have chosen for example $\mathcal{X} = L^q$ with $q < 2$ then $\mathcal{X}^* = L^{q^*}$ with $q^* > 2$. Hence, $J^*_{s^*}$ is differentiable in $\mathcal{X}^*$. Using Newton's method we now have to solve the equation

$$\left[A^* J'_p(A\,x - y^\delta) A\,(J^*_{s^*})'(x^*) + \alpha\, I\right] h^* = -T'_{\alpha,y^\delta}(x), \qquad h^* \in \mathcal{X}^*,$$

for given $x \in \mathcal{X}$. We present the algorithm in detail.

Algorithm 3.4.4

(S0) Choose $x_0 \in \mathcal{X}$, $x_0^* := J_s(x_0)$, *set* $n := 0$.

(S1) If $T'_{\alpha,y^\delta}(x_n) = 0$ *then STOP.*

(S2) Calculate the solution $h_n^* \in \mathcal{X}$ *of the equation*

$$\left[A^* J'_p(A\,x_n - y^\delta)\, A\,(J^*_{s^*})'(x_n^*) + \alpha\, I\right] h^* = -T'_{\alpha,y^\delta}(x_n).$$

(S3) Calculate

$$x_{n+1}^* := x_n^* + h_n^*, \quad \text{and} \quad x_{n+1} := J^*_{s^*}(x_{n+1}^*).$$

(S4) Set $n := n+1$ *and go to step (S1).*

We can present the following local convergence result.

Lemma 3.4.2 *Assume* J_p *and* $J^*_{s^*}$ *to be Gâteaux-differentiable and* A *is supposed to be compact. Then the sequence* $\{x_n\}$ *generated by Algorithm 3.4.4 is well-defined and we have local convergence* $x_n \to x_\alpha^\delta$ *as* $n \to \infty$. *If – additionally – the space* $\mathcal{X}$ *is* r*-smooth for some* $r \in (1,2]$ *then we have convergence at least with a rate*

$$\|x_{n+1} - x_\alpha^\delta\| \le C\, \|x_n - x_\alpha^\delta\|^{\frac{2(r-1)}{s-1}}$$

with constant $C > 0$.

PROOF. We show that $-\alpha < 0$ cannot be an eigenvalue of the operator $B := A^* J'_p(A\,x_n - y^\delta)\, A\,(J^*_{s^*})'(x_n^*) : \mathcal{X}^* \longrightarrow \mathcal{X}^*$. Assume the opposite. Then there exists an element $x^* \in \mathcal{X}^*$ such that $B\,x^* = -\alpha\, x^*$ and $x^* \neq 0$. We multiply the equation with $(J^*_{s^*})'(x_n^*)\, x^* \in \mathcal{X}$. Then the right hand side gives

$$-\alpha\langle x^*, (J^*_{s^*})'(x_n^*)\, x^*\rangle < 0$$

by the positivity of the operator $(J^*_{s^*})'(x^*)$. On the other hand we have

$$\langle B\,x^*, (J^*_{s^*})'(x^*_n)\,x^*\rangle = \langle J'_p(A\,x_n - y^\delta)\,A\,(J^*_{s^*})'(x^*_n)\,x^*, A\,(J^*_{s^*})'(x^*_n)\,x^*\rangle \geq 0$$

by positivity of $J'_p(A\,x_n - y^\delta)$. This contradicts the assumption. By compactness of A the operator B is compact as well and we can apply the Riesz-Schauder theory, see e.g. [55, Section XIII.1], which gives the existence of the (continuous) inverse operator $(B + \alpha\,I)^{-1} : \mathcal{X}^* \longrightarrow \mathcal{X}^*$. Hence, Algorithm 3.4.4 convergences locally in $\mathcal{X}^*$ with quadratic speed of convergence, i.e $\|x^*_{n+1} - J_s(x^\delta_\alpha)\| \leq C_N\|x^*_n - J_s(x^\delta_\alpha)\|^2$ holds for some constant $C_N > 0$. Moreover, by the properties of the spaces we have

$$\begin{aligned}
\|x_{n+1} - x^\delta_\alpha\| &\leq \left(\frac{s}{C_s}\right)^{\frac{1}{s}} \Delta_s(x_{n+1}, x^\delta_\alpha)^{\frac{1}{s}} \\
&= \left(\frac{s}{C_s}\right)^{\frac{1}{s}} \Delta^*_{s^*}(J_s(x^\delta_\alpha), x^*_{n+1})^{\frac{1}{s}} \\
&\leq \left(\frac{s}{C_s}\right)^{\frac{1}{s}} \left(\frac{G^*_{s^*}}{s^*}\right)^{\frac{1}{s}} \|x^*_{n+1} - J_s(x^\delta_\alpha)\|^{\frac{s^*}{s}} \\
&\leq C_N \left(\frac{s}{C_s}\right)^{\frac{1}{s}} \left(\frac{G^*_{s^*}}{s^*}\right)^{\frac{1}{s}} \|x^*_n - J_s(x^\delta_\alpha)\|^{\frac{2}{s-1}} \\
&\leq C_N \left(\frac{(s-1)G^*_{s^*}}{C_s}\right)^{\frac{1}{s}} \left(\frac{G_r}{(r-1)C^*_{r^*}}\right)^{\frac{2}{r^*(s-1)}} \|x_n - x^\delta_\alpha\|^{\frac{2(r-1)}{s-1}}
\end{aligned}$$

by repeating the first steps with respect to r^*-convexity of $\mathcal{X}^*$ and r-smoothness of $\mathcal{X}$. The identity $\Delta_s(x_{n+1}, x^\delta_\alpha) = \Delta^*_{s^*}(J_s(x^\delta_\alpha), x^*_{n+1})$ was already mentioned in Example 2.3.1. ■

Since $r \leq 2$ and $s \geq 2$ we remark that quadratic speed of convergence can be proved only in the Hilbert space case $r = s = 2$.

3.5 Discretized equations

We now discuss the regularization of discretized operator equations. Here, we restrict ourself to

- operators F with compact derivative $F'(x^\dagger)$,
- exponents $p > 1$,
- the case of low order convergence rates and
- the penalty functional P satisfies the coercivity assumption (A7).

Therefore, let $R_h : \mathcal{X} \longrightarrow \mathcal{X}_h$ and $Q_h : \mathcal{Y} \longrightarrow \mathcal{Y}_h$ denote projections onto finite dimensional subspaces $\mathcal{X}_h$ of $\mathcal{X}$ and $\mathcal{Y}_h$ of $\mathcal{Y}$ respectively. As usual, $h > 0$ describes the levels of discretization. Then the discretized equation (3.2) reads as

$$Q_h F(x) = Q_h y^\delta, \qquad x \in \mathcal{X}_h \cap \mathcal{D}(F).$$

Of course, if $\mathcal{R}(P_h)$ and $\mathcal{R}(Q_h)$ are finite-dimensional, at least the linear equation (3.3) is not ill-posed anymore. On the other hand, (3.3) becomes more and more ill-conditioned the smaller we choose h. Finite dimensional approximation in combination with a choice of the discretization level h depending of the noise level δ is well-known as regularization by projection, see [74] and [7, Chapter 7]. We also refer to [52] to some newer results and further references therein. However, numerous numerical studies indicate that we can apply often only a coarse discretization wich is insufficient for problems arising in practical applications. Hence it makes sense to combine regularization and discretization in an appropriate manner.

The Hilbert space situation for linear operator was considered in [38] for arbitrary linear regularization methods.

3.5.1 Discretized linear equations

In a first step we consider linear equations. Therefore we set $A_h := Q_h A\, R_h$. Then we replace the linear equation (3.3) by

$$A_h x = Q_h y^\delta, \qquad x \in \mathcal{X}_h,\ y^\delta \in \mathcal{Y}. \tag{3.46}$$

The discretized Tikhonov regularization approach now reads as

$$\frac{1}{p}\|A_h x - Q_h y^\delta\|^p + \alpha P(x) \to \min \quad \text{subject to} \quad x \in \mathcal{X}_h \cap \mathcal{D}(P). \tag{3.47}$$

A solution of (3.47) we denote with $x^\delta_{\alpha,h}$. Moreover we introduce the notation $x^\dagger_h := R_h x^\dagger$ and suppose $x^\dagger_h \in \mathcal{D}(P)$.

We start to derive estimates for the approximation error $\|x^\delta_{\alpha,h} - x^\dagger\|$. By the minimizing property of $x^\delta_{\alpha,h}$ we conclude

$$\frac{1}{p}\|A_h x^\delta_{\alpha,h} - Q_h y^\delta\|^p + \alpha\left(P(x^\delta_{\alpha,h}) - P(x^\dagger)\right) \le \frac{1}{p}\|A_h x^\dagger_h - Q_h y^\delta\|^p + \alpha\left(P(x^\dagger_h) - P(x^\dagger)\right).$$

The first term of the right hand side we estimate by

$$\begin{aligned}\frac{1}{p}\|A_h x^\dagger_h - Q_h y^\delta\|^p &\le \frac{1}{p}\left\|Q_h\left(A\, R_h x^\dagger - A\, x^\dagger + A\, x^\dagger - y^\delta\right)\right\|^p \\ &\le \frac{2^{p-1}}{p}\left(\|Q_h A(R_h - I)x^\dagger\|^p + \|Q_h(y - y^\delta)\|^p\right).\end{aligned}$$

We now assume that $\xi^\dagger \in \partial P(x^\dagger)$ satisfies the approximative source condition $\xi^\dagger = A^\star\omega + \upsilon$ for some $\omega \in \mathcal{Y}^*$ and $\upsilon \in \mathcal{X}^*$. The we continue with

$$P(x^\delta_{\alpha,h}) - P(x^\dagger) = \langle \xi^\dagger, x^\delta_{\alpha,h} - x^\dagger\rangle + D_{\xi^\dagger}(x^\delta_{\alpha,h}, x^\dagger) = \langle A^\star\omega + \upsilon, x^\delta_{\alpha,h} - x^\dagger\rangle + D_{\xi^\dagger}(x^\delta_{\alpha,h}, x^\dagger)$$

and

$$P(x^\dagger_h) - P(x^\dagger) = \langle \xi^\dagger, x^\dagger_h - x^\dagger\rangle + D_{\xi^\dagger}(x^\dagger_h, x^\dagger) = \langle A^\star\omega + \upsilon, x^\dagger_h - x^\dagger\rangle + D_{\xi^\dagger}(x^\dagger_h, x^\dagger).$$

Combining these estimates we arrive at

$$\begin{aligned}
\frac{1}{p}\|A_h x_{\alpha,h}^\delta - Q_h y^\delta\|^p + \alpha D_{\xi^\dagger}(x_{\alpha,h}^\delta, x^\dagger) &\leq \frac{2^{p-1}}{p}\left(\|Q_h A(R_h - I)x^\dagger\|^p + \|Q_h(y-y^\delta)\|^p\right) \\
&\quad + \alpha D_{\xi^\dagger}(x_h^\dagger, x^\dagger) + \alpha\langle A^\star\omega + v, x_h^\dagger - x_{\alpha,h}^\delta\rangle \\
&\leq \frac{2^{p-1}}{p}\left(\|Q_h A(R_h - I)x^\dagger\|^p + \|Q_h(y-y^\delta)\|^p\right) \\
&\quad + \alpha D_{\xi^\dagger}(x_h^\dagger, x^\dagger) + \alpha\langle \omega, A x_h^\dagger - A x_{\alpha,h}^\delta\rangle \\
&\quad + \alpha\|v\|\|x_h^\dagger - x_{\alpha,h}^\delta\|.
\end{aligned}$$

We further have

$$\begin{aligned}
A x_h^\dagger &= A R_h x^\dagger - A_h x^\dagger + A_h x^\dagger - Q_h A x^\dagger + Q_h A x^\dagger - Q_h y^\delta + Q_h y^\delta \\
&= (I - Q_h) A R_h x^\dagger + Q_h A(R_h - I)x^\dagger + Q_h(y - y^\delta) + Q_h y^\delta
\end{aligned}$$

and

$$A x_{\alpha,h}^\delta = A x_{\alpha,h}^\delta - Q_h A x_{\alpha,h}^\delta + Q_h A x_{\alpha,h}^\delta = (I - Q_h) A x_{\alpha,h}^\delta + A_h x_{\alpha,h}^\delta.$$

This gives

$$\begin{aligned}
\frac{1}{p}\|A_h x_{\alpha,h}^\delta - Q_h y^\delta\|^p + \alpha D_{\xi^\dagger}(x_{\alpha,h}^\delta, x^\dagger) &\leq \frac{2^{p-1}}{p}\left(\|Q_h A(R_h - I)x^\dagger\|^p + \|Q_h(y-y^\delta)\|^p\right) \\
&\quad + \alpha D_{\xi^\dagger}(x_h^\dagger, x^\dagger) + \alpha\|v\|\|R_h(x^\dagger - x_{\alpha,h}^\delta)\| \\
&\quad + \alpha\|\omega\|\|(I - Q_h) A R_h(x^\dagger - x_{\alpha,h}^\delta)\| \\
&\quad + \alpha\|\omega\|\|Q_h A(R_h - I)x^\dagger\| \\
&\quad + \alpha\|\omega\|\left(\|Q_h(y-y^\delta)\| + \|Q_h y^\delta - A_h x_{\alpha,h}^\delta\|\right).
\end{aligned}$$

We continue as in Section 3.2.4 by applying Young's inequality three times to the last terms of the right hand side. Under Assumption (A7) we get

$$\begin{aligned}
\|x_{\alpha,h}^\delta - x^\dagger\|^s &\leq \frac{1}{C_s} D_{\xi^\dagger}(x_{\alpha,h}^\delta, x^\dagger) \\
&\leq \frac{2^{p-1}+1}{pC_s}\left(\frac{\|Q_h A(R_h - I)x^\dagger\|^p}{\alpha} + \frac{\|Q_h(y-y^\delta)\|^p}{\alpha}\right) \\
&\quad + \frac{1}{C_s} D_{\xi^\dagger}(x_h^\dagger, x^\dagger) + \frac{3(p-1)}{C_s p}\alpha^{\frac{1}{p-1}}\|\omega\|^{\frac{p}{p-1}} \\
&\quad + \frac{\|v\|}{C_s}\|R_h(x^\dagger - x_{\alpha,h}^\delta)\| + \frac{1}{C_s}\|\omega\|\|(I - Q_h) A R_h(x^\dagger - x_{\alpha,h}^\delta)\|
\end{aligned}$$

We now suppose $v \neq 0$, i.e. $\xi^\dagger \notin \mathcal{R}(A^\star)$. Then the modified version of Young's equality with $\varepsilon = \frac{1}{2s}$ now yields

$$\frac{\|v\|}{C_s}\|R_h(x^\dagger - x_{\alpha,h}^\delta)\| \leq \frac{s-1}{s} 2^{\frac{1}{s-1}}\left(\frac{\|v\|\|R_h\|}{C_s}\right)^{\frac{s}{s-1}} + \frac{1}{2s}\|x_{\alpha,h}^\delta - x^\dagger\|^s$$

and

$$\frac{\|\omega\|}{C_s}\|(I-Q_h)A\,R_h(x^\dagger - x^\delta_{\alpha,h})\| \le \frac{s-1}{s}2^{\frac{1}{s-1}}\left(\frac{\|\omega\|\|(I-Q_h)A\|\|R_h\|}{C_s}\right)^{\frac{s}{s-1}} + \frac{1}{2s}\|x^\delta_{\alpha,h} - x^\dagger\|^s.$$

We finally get the estimate

$$\begin{aligned}\|x^\delta_{\alpha,h} - x^\dagger\|^s &\le \mathcal{C}_1\left(\frac{\|A(R_h - I)x^\dagger\|^p}{\alpha} + \frac{\delta^p}{\alpha}\right) + \frac{1}{C_s}D_{\xi^\dagger}(x_h^\dagger, x^\dagger)\\ &\quad + \mathcal{C}_2\alpha^{\frac{1}{p-1}}\|\omega\|^{\frac{p}{p-1}} + \mathcal{C}_3\left(\|\upsilon\|^{\frac{s}{s-1}} + (\|\omega\|\,\|(I-Q_h)A\|)^{\frac{s}{s-1}}\right)\end{aligned}$$

with constants

$$\mathcal{C}_1 := \frac{s(2^{p-1}+1)}{(s-1)pC_s}\|Q_h\|^p, \quad \mathcal{C}_2 := \frac{3s(p-1)}{C_s(s-1)p} \quad \text{and} \quad \mathcal{C}_3 := \frac{2^{\frac{1}{s-1}}}{C_s^{\frac{s}{s-1}}}\|R_h\|^{\frac{s}{s-1}}.$$

On the other hand, if $\xi^\dagger \in \mathcal{R}(A^\star)$ we can suppose $\upsilon = 0$. Then a similar calculation leads to the estimate

$$\begin{aligned}\|x^\delta_{\alpha,h} - x^\dagger\|^s &\le \mathcal{C}_1\left(\frac{\|A(R_h - I)x^\dagger\|^p}{\alpha} + \frac{\delta^p}{\alpha}\right) + \frac{1}{C_s}D_{\xi^\dagger}(x_h^\dagger, x^\dagger)\\ &\quad + \mathcal{C}_2\alpha^{\frac{1}{p-1}}\|\omega\|^{\frac{p}{p-1}} + \tilde{\mathcal{C}}_3\left(\|\omega\|\,\|(I-Q_h)A\|\right)^{\frac{s}{s-1}}\end{aligned}$$

with

$$\tilde{\mathcal{C}}_3 := \left(\frac{\|R_h\|}{C_s}\right)^{\frac{s}{s-1}}.$$

We now introduce quantifications for the discretization levels. Therefore we assume

$$\|A(I - R_h)\| \le \xi_h \quad \text{and} \quad \|(I-Q_h)A\| \le \eta_h$$

for given bounds $\xi_h > 0$ and $\eta_h > 0$ and $\|R_h\| \le C_R$ and $\|Q_h\| \le C_Q$. We suppose $\xi_h \to 0$ and $\eta_h \to 0$ as $h \to 0$. This approach of definition for the bounds of the discretization error has been well-established for discretized (linear) inverse problems in Hilbert spaces, see e.g. [80] and [68]. It turns out in Hilbert spaces that these numbers are sufficient for describing a proper choice of the discretization level depending on the regularization parameter α and/or the noise level δ. A proper choice of the discretization means that the discretization errors are of the same order than the regularization error and/or the perturbation error. This observation does not hold true in Banach spaces. Here, the additional term $D_{\xi^\dagger}(x_h^\dagger, x^\dagger)$ occurs which cannot be estimated by the discretization bounds ξ_h and η_h without additional assumptions. So we need a further condition to get an estimate for the Bregman distance $D_{\xi^\dagger}(x_h^\dagger, x^\dagger)$. We cite the following result.

Theorem 3.5.1 *Assume (A1)-(A5) and (A7). For two constants $C_\xi, C_\eta > 0$ we set $\xi_h \le C_\xi\alpha^{\frac{1}{p-1}}$, $\eta_h \le C_\eta\alpha^{\frac{1}{p}}$ if $s \le p$ and $\eta_h \le C_\eta\alpha^{\frac{s-1}{s(p-1)}}$ for $s > p$. Additionally we suppose $D_{\xi^\dagger}(x_h^\dagger, x^\dagger) \le C_h\alpha^{\frac{1}{p-1}}$ for some constant $C_h > 0$.*

(i) If $\xi^\dagger \in \mathcal{R}(A^\star)$, then the estimate

$$\|x_{\alpha,h}^\delta - x^\dagger\| \leq D_1 \frac{\delta^{\frac{p}{s}}}{\alpha^{\frac{1}{s}}} + D_2 \alpha^{\frac{1}{p-1}} \tag{3.48}$$

holds for two constants $D_1, D_2 > 0$.

(ii) If $\xi^\dagger \in \overline{\mathcal{R}(A^\star)} \setminus \mathcal{R}(A^\star)$ with distance function $d(R) = d(R;\xi^\dagger)$ then we have the estimate

$$\|x_{\alpha,h}^\delta - x^\dagger\| \leq \tilde{D}_1 \frac{\delta^{\frac{p}{s}}}{\alpha^{\frac{1}{s}}} + \tilde{D}_2 d\left(\Theta^{-1}(\alpha)\right)^{\frac{1}{s-1}} \tag{3.49}$$

with two constants $\tilde{D}_1, \tilde{D}_2 > 0$. Here $\Theta(R) := d(R)^{\frac{s(p-1)}{s-1}} R^{-p}$ is the function of Theorem 3.2.3 with $\kappa = c_1 = 1$.

PROOF. We first deal with the second part. For $R > 0$ we have $\xi^\dagger = A^\dagger\omega + \upsilon$ with $\|\omega\| \leq R$ and $\|\upsilon\| \leq d(R)$. Using the bounds of the discretization we arrive at

$$\begin{aligned}\|x_{\alpha,h}^\delta - x^\dagger\|^s \leq & \left(\mathcal{C}_1 C_\xi^p \|x^\dagger\|^p + \frac{C_h}{C_s}\right)\alpha^{\frac{1}{p-1}} + \mathcal{C}_1 \frac{\delta^p}{\alpha} \\ & + \mathcal{C}_2 \alpha^{\frac{1}{p-1}} R^{\frac{p}{p-1}} + C_3\left(d(R) + R^{\frac{s}{s-1}} \eta_h^{\frac{s}{s-1}}\right).\end{aligned}$$

We set $R := \Theta^{-1}(\alpha)$ such that

$$\alpha^{\frac{1}{p-1}} R^{\frac{p}{p-1}} = \alpha^{\frac{1}{p-1}} \left[\Theta^{-1}(\alpha)\right]^{\frac{p}{p-1}} = d\left(\Theta^{-1}(\alpha)\right)^{\frac{s}{s-1}}.$$

Moreover, with $R_{min} := \Theta^{-1}(\alpha_{max})$ we have

$$\alpha^{\frac{1}{p-1}} = \alpha^{\frac{1}{p-1}} \Theta^{-1}(\alpha)^{\frac{p}{p-1}} \Theta^{-1}(\alpha)^{-\frac{p}{p-1}} \leq d\left(\Theta^{-1}(\alpha)\right)^{\frac{s}{s-1}} R_{min}^{-\frac{p}{p-1}} \tag{3.50}$$

using that the function $\Theta^{-1}(\alpha)$ is strictly decreasing. We now assume $s \leq p$. Then we conclude $\frac{s}{s-1} \geq \frac{p}{p-1}$ and hence

$$R^{\frac{s}{s-1}} \eta_h^{\frac{s}{s-1}} \leq C_\eta^{\frac{s}{s-1}} \left(R^{\frac{p}{p-1}} \alpha^{\frac{1}{p-1}}\right)^{1+\frac{p-s}{p(s-1)}} \leq C_\eta^{\frac{s}{s-1}} d\left(\Theta^{-1}(\alpha)\right)^{\frac{s}{s-1}} d\left(R_{min}\right)^{\frac{s(p-s)}{p(s-1)^2}}.$$

Hence we obtained estimate (3.49) with $\tilde{D}_1 := \mathcal{C}_1^{\frac{1}{s}}$ and

$$\tilde{D}_2 := \left[\left(\mathcal{C}_1 C_\xi^p \|\xi^\dagger\|^p + \frac{C_h}{C_s}\right) R_{min}^{-\frac{p}{p-1}} + \mathcal{C}_2 + C_3\left(1 + C_\eta^{\frac{s}{s-1}} d\left(R_{min}\right)^{\frac{s(p-s)}{p(s-1)^2}}\right)\right]^{\frac{1}{s}}.$$

We now suppose $s > p$. Then we can apply Young's inequality with $p_1 = \frac{p}{p-1}\frac{s-1}{s}$ and $p_2 = \frac{p(s-1)}{s-p} > 1$ to obtain

$$\begin{aligned}\left(\alpha^{\frac{s}{p(s-1)}} R^{\frac{s}{s-1}}\right)\left(\frac{\eta_h^{\frac{s}{s-1}}}{\alpha^{\frac{s}{p(s-1)}}}\right) & \leq \frac{s(p-1)}{(s-1)p} \alpha^{\frac{1}{p-1}} R^{\frac{p}{p-1}} + \frac{s-p}{p(s-1)} \frac{\eta_h^{\frac{sp}{s-p}}}{\alpha^{\frac{s}{s-p}}} \\ & \leq \frac{s(p-1)}{(s-1)p} d\left(\Theta^{-1}(\alpha)\right)^{\frac{s}{s-1}} + \frac{s-p}{p(s-1)} C_\eta^{\frac{sp}{s-p}} \alpha^{\frac{1}{p-1}}.\end{aligned}$$

We use again (3.50) for proving the estimate (3.49) where we now set

$$\tilde{D}_2 := \left[\left(\mathcal{C}_1 C_\xi^p + \frac{C_h}{C_s}\right) R_{min}^{-\frac{p}{p-1}} + \mathcal{C}_2 + \frac{s(p-1)}{(s-1)p} + \mathcal{C}_3\left(1 + \frac{s-p}{p(s-1)} C_\eta^{\frac{sp}{s-p}} R_{min}^{-\frac{p}{p-1}}\right)\right]^{\frac{1}{s}}.$$

This proves the second part. If $\xi^\dagger \in \mathcal{R}(A^\star)$, i.e. $\xi^\dagger = A^\star\omega$ for some $\omega \in \mathcal{Y}^*$ we derive

$$\|x_{\alpha,h}^\delta - x^\dagger\|^s \le \left(\mathcal{C}_1 C_\xi^p \|x^\dagger\|^p + \frac{C_h}{C_s} + \mathcal{C}_2\|\omega\|^{\frac{p}{p-1}}\right)\alpha^{\frac{1}{p-1}} + \mathcal{C}_1\frac{\delta^p}{\alpha} + \tilde{\mathcal{C}}_3\|\omega\|^{\frac{s}{s-1}}\eta_h^{\frac{s}{s-1}}.$$

For $s \le p$ we have $\alpha^{\frac{s}{s-1}} \le \alpha_{max}^\nu \alpha^{\frac{p}{p-1}}$ with $\nu := \frac{s}{s-1} - \frac{p}{p-1} = \frac{p-s}{(s-1)(p-1)} > 0$ and hence

$$\eta_h^{\frac{s}{s-1}} \le C_\eta^{\frac{s}{s-1}} \alpha^{\frac{s}{p(s-1)}} \le C_\eta^{\frac{s}{s-1}} \alpha_{max}^{\frac{\nu}{p}} \alpha^{\frac{1}{p-1}}.$$

On the other hand, for $s > p$ we have $\eta_h^{\frac{s}{s-1}} \le C_\eta^{\frac{s}{s-1}} \alpha^{\frac{1}{p-1}}$. This shows the estimate (3.48) with $D_1 := \mathcal{C}_1^{\frac{1}{s}}$ and

$$D_2 := \left[\mathcal{C}_1 C_\xi^p \|x^\dagger\|^p + \frac{C_h}{C_s} + \mathcal{C}_2\|\omega\|^{\frac{p}{p-1}} + \tilde{\mathcal{C}}_3\left(C_\eta\|\omega\|\right)^{\frac{s}{s-1}} \alpha_{max}^{\hat{\nu}}\right]^{\frac{1}{s}}$$

with $\hat{\nu} := [p(p-1)(s-1)]^{-1}\max\{p-s, 0\}$. The proof is complete. ■

The estimates (3.48) and (3.49) show that the discretization strategy suggested in Theorem 3.5.1 does not destroy the achieved convergence rates of Section 3.2.4 for the undiscretized linear equation (3.3). In particular, under the presented a-priori parameter choices of Theorem 3.2.1 for fulfilled source condition (3.11) and Theorem 3.2.3 in connection with the approximate source condition (3.12) and underlying distance function we derive that the errors $\|x_\alpha^\delta - x^\dagger\|$ and $\|x_{\alpha,h}^\delta - x^\dagger\|$ are of same order with respect to the noise level δ. The corresponding result is presented below.

Corollary 3.5.1 *Let the conditions of Theorem 3.5.1 hold.*

(i) If $\xi^\dagger \in \mathcal{R}(A^\star)$ then an a-priori parameter choice $\alpha \sim \delta^{p-1}$ leads to the convergence rate

$$\|x_{\alpha,h}^\delta - x^\dagger\| = \mathcal{O}\left(\delta^{\frac{1}{s}}\right) \quad as \quad \delta \to 0.$$

(ii) Let $\xi^\dagger \in \overline{\mathcal{R}(A^\star)} \setminus \mathcal{R}(A^\star)$ has distance function $d(R) = d(R;\xi^\dagger)$. We define the functions $\Theta(R)$, $\Psi(\alpha)$ and $\Phi(R)$ as in Theorem 3.2.3 with $\kappa = c_1 = 1$. Then an a-priori parameter choice $\alpha := \Psi^{-1}(\delta)$ leads to the convergence rate

$$\|x_{\alpha,h}^\delta - x^\dagger\| = \mathcal{O}\left(d\left(\Phi^{-1}(\delta)\right)^{\frac{1}{s-1}}\right) \quad as \quad \delta \to 0.$$

Remark 3.5.1 *Of practical interest are the applicability of a-posteriori choices of the regularization parameter α without knowledge of the distance functions $d(R;\xi^\dagger)$. The structure of the estimates (3.48) and (3.49) of Theorem 3.5.1 shows that we can apply the balancing principle for a proper choice of the regularization parameter α keeping the optimal convergence rates of Corollary 3.5.1.*

3.5.2 Extension to nonlinear problems

We discuss the generalization to nonlinear equations. We have to assume the following

- For all $h > 0$ we have with $L < 1$

$$\|Q_h(F(x) - F(x^\dagger) - F'(x^\dagger)(x - x^\dagger))\| \le L\,\|Q_h(F(x) - F(x^\dagger))\|$$

for all $x \in \mathcal{B}_{\varrho_x}(x^\dagger) \cap \mathcal{X}_h \cap \mathcal{D}$.

With $L < 1$ we derive

$$\begin{aligned}\|Q_h(F(x) - F(x^\dagger))\| &\le \|Q_h(F(x) - F(x^\dagger) - F'(x^\dagger)(x - x^\dagger))\| + \|Q_h(F'(x^\dagger)(x - x^\dagger))\| \\ &\le L\,\|Q_h(F(x) - F(x^\dagger))\| + \|Q_h(F'(x^\dagger)(x - x^\dagger))\|\end{aligned}$$

and hence

$$\|Q_h(F(x) - F(x^\dagger))\| \le \frac{1}{1-L}\|Q_h(F'(x^\dagger)(x - x^\dagger))\|.$$

Now we continue with

$$\begin{aligned}\frac{1}{p}\|Q_h(F(x_h^\dagger) - y^\delta)\|^p &= \frac{1}{p}\|Q_h(F(x_h^\dagger) - F(x^\dagger) + y - y^\delta)\|^p \\ &\le \frac{2^{p-1}}{p}\left(\|Q_h(F(x_h^\dagger) - F(x^\dagger))\|^p + \|Q_h(y - y^\delta)\|^p\right) \\ &\le \frac{2^{p-1}}{p(1-L)}\|[Q_hF'(x^\dagger)(R_h - I)]x^\dagger\|^p + \frac{2^{p-1}}{p}\|Q_h(y - y^\delta)\|^p.\end{aligned}$$

Moreover, we estimate

$$\begin{aligned}&\left\|F'(x^\dagger)x_h^\dagger - F'(x^\dagger)x_{\alpha,h}^\delta\right\| \\ &\quad = \big\|(I - Q_h)F'(x^\dagger)R_hx^\dagger + Q_hF'(x^\dagger)(R_h - I)x^\dagger + Q_hF'(x^\dagger)x^\dagger \\ &\qquad\qquad -(I - Q_n)F'(x^\dagger)x_{\alpha,h}^\delta + Q_nF'(x^\dagger)x_{\alpha,h}^\delta\big\| \\ &\quad \le \big\|(I - Q_h)F'(x^\dagger)R_h(x^\dagger - x_{\alpha,h}^\delta)\big\| + \big\|Q_hF'(x^\dagger)(R_h - I)x^\dagger\big\| \\ &\qquad + \big\|Q_n(F'(x^\dagger)(x_{\alpha,h}^\delta - x^\dagger) + F(x_{\alpha,h}^\delta) - F(x^\dagger))\big\| + \big\|Q_n(F(x_{\alpha,h}^\delta) - F(x^\dagger))\big\| \\ &\quad \le \big\|(I - Q_h)F'(x^\dagger)R_h(x^\dagger - x_{\alpha,h}^\delta)\big\| + \big\|Q_hF'(x^\dagger)(R_h - I)x^\dagger\big\| \\ &\qquad +(1 + L)\left(\big\|Q_h(F(x_{\alpha,h}^\delta) - y^\delta)\big\| + \big\|Q_h(y^\delta - y)\big\|\right)\end{aligned}$$

These calculations show that we can extend the error estimates and convergence rates results of Theorem 3.5.1 and Corollary 3.5.1 as long as the above nonlinearity restriction is fulfilled.

3.6 Numerical examples

We again deal with Example 1.2.1. In particular we have $\mathcal{X} := L^q(0,1)$, $\mathcal{Y} := L^p(0,1)$, $1 < p, q < \infty$ and $P(x) := \frac{1}{q}\|x\|^q$.

3.6.1 Theoretical results

In order to compare numerical results with the derived convergence rate theory we apply these theoretical statements to a specific example. Therefore we assume $x^\dagger \equiv 1$. We distinguish between the low order and high order situation. From the theoretical results we now want to propose convergence rates for given exact data as well as for given noisy data. In particular, we want to derive exponents $\nu_\alpha > 0$ and $\nu_\delta > 0$ such that the estimates

$$\|x_\alpha - x^\dagger\| \le C_1 \alpha^{\nu_\alpha} \quad \text{and} \quad \|x_\alpha^\delta - x^\dagger\| \le C_2 \delta^{\nu_\delta}$$

hold for two constants $C_1, C_2 > 0$ when in the noisy case the regularization parameter $\alpha = \alpha(\delta)$ is chosen as in the a-priori choice of 3.2.3 suggested.

a) The low-order case

In the low order case we derived for $\xi^\dagger := J_q(x^\dagger) \equiv 1$ the distance function

$$d(R) := d(R;\xi^\dagger) \le \mathcal{C} R^{\frac{1-q}{q}p}, \qquad R \ge 1,$$

see Example 3.2.1. Throughout the section let $\mathcal{C} > 0$ be a generic constant. Moreover by the convexity of the space $\mathcal{X} = L^q(0,1)$ we have $s := \max\{2, q\}$ in assumption (A7). Then, from Theorem 3.2.3 with $c_1 = \kappa = 1$ we conclude

$$\Phi(R) := d(R)^{\frac{s}{s-1}} R^{-1}, \qquad R > 0,$$

which leads to the convergence rates

$$D_{\xi^\dagger}(x_\alpha^\delta, x^\dagger) \le \mathcal{C}\, d\left(\Phi^{-1}(\delta)\right)^{\frac{s}{s-1}} \quad \text{and} \quad \|x_\alpha^\delta - x^\dagger\| \le \mathcal{C}\, d\left(\Phi^{-1}(\delta)\right)^{\frac{1}{s-1}}.$$

We set $\nu := \frac{q-1}{q}p$. Then

$$\Phi(R) = d(R)^{\frac{s}{s-1}} R^{-1} \le \mathcal{C}\, R^{-\frac{s\nu+s-1}{s-1}} = \mathcal{C}\, R^{-\frac{s(\nu+1)-1}{s-1}}$$

holds. This gives

$$\begin{aligned} \|x_\alpha^\delta - x^\dagger\| \le \mathcal{C}\, \delta^{\frac{\nu}{s(\nu+1)-1}} &= \mathcal{C}\, \delta^{\frac{(q-1)p}{s((q-1)p+q)-q}} \\ &= \mathcal{C}\, \delta^{\frac{(q-1)p}{s(q-1)p+q(s-1)}} =: \mathcal{C}\, \delta^{\nu_\delta}. \end{aligned}$$

We also want to examine the noiseless case $\delta = 0$. Again, from the proof of Theorem 3.2.3 with $c_1 = \kappa = 1$ we see that

$$\Theta(R) := d(R)^{\frac{s(p-1)}{s-1}} R^{-p}, \qquad R > 0,$$

which holds the rates

$$D_{\xi^\dagger}(x_\alpha, x^\dagger) \le \mathcal{C}\, d\left(\Theta^{-1}(\alpha)\right)^{\frac{s}{s-1}} \quad \text{and} \quad \|x_\alpha - x^\dagger\| \le \mathcal{C}\, d\left(\Theta^{-1}(\alpha)\right)^{\frac{1}{s-1}}.$$

Here we estimate

$$\Theta(R) = d(R)^{\frac{s(p-1)}{s-1}} R^{-p} \le \mathcal{C}\, R^{-\frac{s\nu(p-1)+p(s-1)}{s-1}}$$

and consequently

$$\begin{aligned}\|x_\alpha - x^\dagger\| \le \mathcal{C}\,\alpha^{\frac{\nu}{s\nu(p-1)+p(s-1)}} &= \mathcal{C}\,\alpha^{\frac{(q-1)p}{s(q-1)p(p-1)+pq(s-1)}}\\ &= \mathcal{C}\,\alpha^{\frac{sp}{s(p-1)+(s-1)\frac{q}{q-1}}} =: \mathcal{C}\,\alpha^{\nu_\alpha}.\end{aligned}$$

We have to differentiate between $q \le 2$ and $q \ge 2$. For $q = 2$ both cases should coincide.

- CASE $q \le 2$. Then $s = 2$ holds. In the noiseless situation we have

$$\nu_\alpha = \frac{2}{2(p-1) + \frac{q}{q-1}}.$$

For $q \to 2$ we have $\nu_\alpha \to \frac{1}{2p}$. In particular, for $p = 2$ we have $\nu_\alpha = \frac{1}{4}$ which is the expected rate in the Hilbert space case. On the other hand for $q \to 1$ we observe $\nu_\alpha \to 0$. That means that the convergence rate slows down to zero for $q \to 1$. This is an immediate consequence of the following observation: in the limit case $\mathcal{X} = L^1(0,1)$ and $\mathcal{X}^* = L^\infty(0,1)$ we see $d(R;\xi^\dagger) \equiv 1$, $R \ge 0$, since $\xi^\dagger \notin \overline{\mathcal{R}(A^\star)}$ (we always have $\omega(1) = 0$ for all $\omega \in \mathcal{R}(A^\star)$).

In the case of noisy data we have

$$\nu_\delta = \frac{p}{2p + \frac{q}{q-1}}.$$

We observe $\nu_\delta \to \frac{p}{2p+2}$ for $q \to 2$ and $\nu_\delta \to 0$ for $q \to 1$. Moreover, for the Hilbert space case $p = q = 2$ we have $\nu_\delta = \frac{1}{3}$ which coincides with known results.

- CASE $q \ge 2$. Then $s = q$ follows. For noiseless data this gives

$$\nu_\alpha = \frac{1}{q(p-1)+q} = \frac{1}{qp}.$$

For $q \to 2$ we again arrive at $\nu_\alpha \to \frac{1}{2p}$ which coincides with the first case. For $q \to \infty$ we obtain $\nu_\alpha \to 0$. The latter case is devoted the weak coercivity condition (A7) on the penalty functional for $q \to \infty$.

For noisy date we observe

$$\nu_\delta = \frac{p}{qp+q} = \frac{p}{(1+p)q}.$$

Here we have for the exponent $\frac{p}{(1+p)q} \to 0$ as $q \to \infty$ and $\frac{p}{(1+p)q} \to \frac{p}{2p+2}$ as $q \to 2$. The latter situation is exactly the same as in the case $q \le 2$ and $q \to 2$.

b) The high-order case

Although even the low order source condition is violated we can deal with the high order reference source condition. Following Example 3.3.7 we have

$$d_p(R) := d_p(R;\xi^\dagger) \le \mathcal{C}\,R^{-\frac{q-1}{p(q-1)+1}}, \qquad R \ge 1.$$

Hence, we now set $\nu := \frac{q-1}{p(q-1)+1}$. We start with the noiseless case. Then from the proof of Theorem 3.3.4 we conclude with $\Theta(R) := \Theta_p(R)$ that

$$\Theta(R) := d_p(R)^{\frac{s(p-1)}{r(s-1)}} R^{-1}, \qquad R > 0,$$

which provides the convergence rates

$$D_{\xi^\dagger}(x_\alpha, x^\dagger) \le \mathcal{C}\, d_p\left(\Theta^{-1}(\alpha)\right)^{\frac{s}{s-1}} \quad \text{and} \quad \|x_\alpha - x^\dagger\| \le \mathcal{C}\, d_p\left(\Theta^{-1}(\alpha)\right)^{\frac{1}{s-1}}.$$

Here $r > 1$ describes the smoothness of the penalty functional P, i.e. we have $r := \min\{2, q\}$. Using the specific power-type structure of the distance function we arrive at

$$\Theta(R) = d_p(R)^{\frac{s(p-1)}{r(s-1)}} R^{-1} \le \mathcal{C}\, R^{-\frac{s\nu(p-1)+r(s-1)}{r(s-1)}}$$

and finally

$$\|x_\alpha - x^\dagger\| \le \mathcal{C}\, \alpha^{\frac{r(s-1)}{s\nu(p-1)+r(s-1)} \frac{\nu}{s-1}} = \mathcal{C}\, \alpha^{\frac{\nu r}{s\nu(p1-)+r(s-1)}} =: \mathcal{C}\, \alpha^{\nu_\alpha}.$$

Again we have to distinguish two situations.

- Case $q \le 2$. Here, $s := 2$ and $r := q$ holds. This gives

$$\nu_\alpha = \frac{q\nu}{2\nu(p-1)+q} = \frac{q(q-1)}{2(q-1)(p-1)+q(p(q-1)+1)} = \frac{q}{2(p-1)+qp+\frac{q}{q-1}}.$$

  Again we discuss the limit situations. For $q \to 2$ we end at $\nu_\alpha \to \frac{2}{2(p-1)+2p+2} = \frac{1}{2p}$ which is the same as in the low order case. For $q \to 1$ we also have $\nu_\alpha \to 0$, i.e. the convergence becomes very slow.

- Case $q \ge 2$. Now $s := q$ and $r := 2$ holds. Here we continue with

$$\begin{aligned} \nu_\alpha = \frac{2\nu}{q\nu(p-1)+2(q-1)} &= \frac{2(q-1)}{q(q-1)(p-1)+2(q-1)(p(q-1)+1)} \\ &= \frac{2}{q(p-1)+2(p(q-1)+1)}. \end{aligned}$$

  Again, for $q \to 2$ we have the expected case $\nu_\alpha \to \frac{1}{2p}$ whereas $\nu_\alpha \to 0$ as $q \to \infty$.

We now examine the situation for given noisy data. Then, from Theorem 3.3.4 we observe with $\Phi(R) := \Phi_p(R)$ that

$$\Phi(R) := \begin{cases} d(R)^{\frac{s(r+p-1)}{(s-1)rp}} R^{-\frac{1}{p}}, & p < 2, \\ d(R)^{\frac{s(r+1)}{2r(s-1)}} R^{-\frac{1}{2}}, & p \ge 2, \end{cases}$$

whereas the convergence rates are of similar structure as in the low order case. We set $t := \min\{p, 2\}$. Using the specific distance function we derive

$$\Phi(R) = d(R)^{\frac{s(r+t-1)}{(s-1)rt}} R^{-\frac{1}{t}} \le \mathcal{C}\, R^{-\frac{\nu s(r+t-1)+r(s-1)}{(s-1)rt}}$$

and finally

$$\|x_\alpha^\delta - x^\dagger\| \le \mathcal{C}\, \delta^{\frac{\nu r t}{\nu s(r+t-1)+r(s-1)}} =: \mathcal{C}\, \delta^{\nu_\delta}.$$

We now have to distinguish between four parameter constellations.

- CASE $q \leq 2$, $p \leq 2$. Here, $t := p$, $s := 2$ and $r := q$ holds. This gives

$$\nu_\delta = \frac{\nu q p}{2\nu(q+p-1)+q} = \frac{(q-1)qp}{2(q-1)(q+p-1)+q(p(q-1)+1)} = \frac{qp}{2(q+p-1)+qp+\frac{q}{q-1}}.$$

For $p = 2$ we have

$$\nu_\delta = \frac{2q}{2(q+1)+2q+\frac{q}{q-1}} = \frac{2q}{4q+2+\frac{q}{q-1}}.$$

Moreover for $q \to 2$ we derive $\nu_\delta \to \frac{1}{3}$ which coincides with the low order convergence rate result. On the other hand, for $q \to 1$ we see that $\nu_\delta \to 0$.

- CASE $q \geq 2$, $p \leq 2$. Now we have $t := p$, $s := q$ and $r := 2$. Then we derive

$$\nu_\delta = \frac{2\nu p}{\nu q(p+1)+2(q-1)} = \frac{2(q-1)p}{q(q-1)(p+1)+2(q-1)(p(q-1)+1)} = \frac{2p}{q(p+1)+2(p(q-1)+1)}.$$

For $p = 2$ we again have $\nu_\delta = \frac{2q}{4q+2+\frac{q}{q-1}}$. Moreover, for $q \to 2$ we arrive at $\nu_\delta \to \frac{1}{3}$ and for $q \to \infty$ we again observe $\nu_\delta \to 0$.

- CASE $q \leq 2$, $p \geq 2$. This gives $s = 2$ and $r = q$ again and $t := 2$. Now we have

$$\nu_\delta = \frac{2\nu q}{2\nu(q+1)+q} = \frac{2(q-1)q}{2(q-1)(q+1)+q(p(q-1)+1)} = \frac{2q}{2(q+1)+qp+\frac{q}{q-1}}.$$

Again we set $p = 2$ which leads to

$$\nu_\delta = \frac{4}{3q+2(2(q-1)+1)} = \frac{4}{7q-2}.$$

The case $q = 2$ again gives $\nu_\delta = \frac{1}{3}$. For $q \to 1$ we observe $\nu_\delta \to 0$ as well.

- CASE $q \geq 2$, $p \geq 2$. We again have $s = q$ and $r = 2$ with $t := 2$. This finally leads to

$$\nu_\delta = \frac{4\nu}{3\nu q+2(q-1)} = \frac{4(q-1)}{3q(q-1)+2(q-1)(p(q-1)+1)} = \frac{4}{3q+2(p(q-1)+1)}.$$

For $p = 2$ we end at $\nu_\delta = \frac{4}{7q-2}$ again. So we only have to consider the limit case $q \to \infty$ which also gives $\nu_\delta \to 0$.

	$q \leq 2$	$q \geq 2$
ν_α	$\dfrac{1}{2(p-1)+\frac{q}{q-1}}$	$\dfrac{1}{pq}$
ν_δ	$\dfrac{p}{2p+\frac{q}{q-1}}$	$\dfrac{p}{(1+p)q}$

Table 3.1: Theoretical convergence rates in the low-order case

		$p \leq 2$	$p \geq 2$
$q \leq 2$	ν_α	$\dfrac{q}{2(p-1)+qp+\frac{q}{q-1}}$	
	ν_δ	$\dfrac{qp}{2(q+p-1)+pq+\frac{q}{q-1}}$	$\dfrac{2q}{2(q+1)+pq+\frac{q}{q-1}}$
$q \geq 2$	ν_α	$\dfrac{2}{q(p-1)+2(p(q-1)+1)}$	
	ν_δ	$\dfrac{2p}{q(p+1)+2(p(q-1)+1)}$	$\dfrac{4}{3q+2(p(q-1)+1)}$

Table 3.2: Theoretical convergence rates in the high-order case

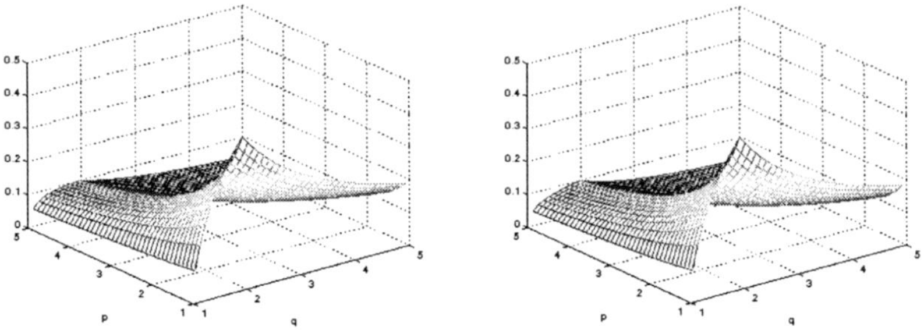

Figure 3.1: Theoretical convergence rates for given exact data in the low-order case (left) and the high-order case (right)

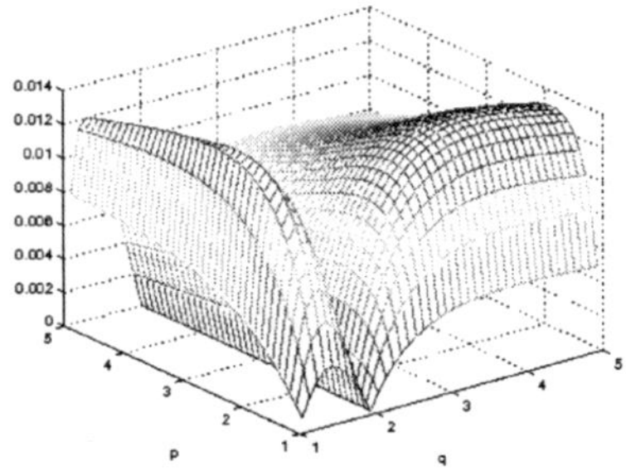

Figure 3.2: Difference of the theoretical convergence rates for given exact data

We summarize these results in Table 3.1 for the low-order case and in Table 3.2 for the high-order situation.

We also present a graphical illustration of the theoretical dependence of the convergence rates on the given parameter q and p. Figure 3.1 shows the rates ν_α for the low-order case as well as for the high-order estimates. Moreover, Figure 3.2 shows the difference between low-order and high-order results. Since the difference is always non-negative we observe that the low-order case in general propose better rates than the high-order case. This it not surprisingly since additional assumptions have to be made in the high-order analysis. For $q = 2$ both approaches provides the same rates.

Similar presentations can be found in Figure 3.3 and Figure 3.4 for the case of given noisy data. Here again the low-order calculation provide in general better rates. Only in the Hilbert space situation $p = q = 2$ both approaches coincides.

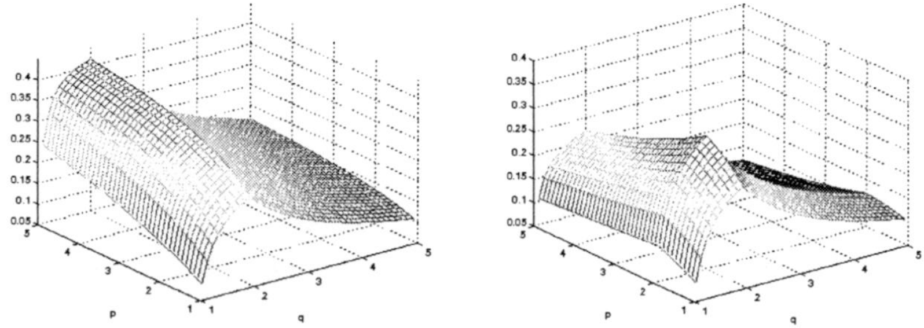

Figure 3.3: Theoretical convergence rates for given noisy data in the low-order case (left) and the high-order case (right)

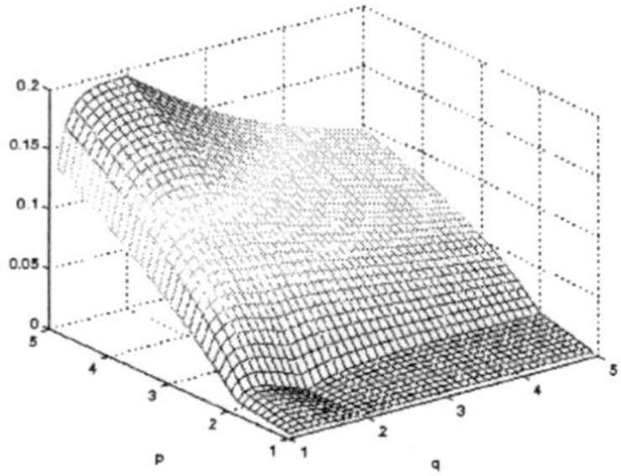

Figure 3.4: Difference of the theoretical convergence rates for given noisy data

3.6.2 The sample functions and discretization

Throughout the numerical studies we will us the following three sample functions:

$$\begin{aligned} x_1^\dagger(t) &:= \left(t-\frac{1}{2}\right)^2+\frac{1}{10}, && t\in[0,1], \\ x_2^\dagger(t) &:= \left(3\left(t-\frac{3}{10}\right)^2+\frac{1}{10}\right)(1-t), && t\in[0,1], \qquad \text{and} \\ x_3^\dagger(t) &:= \frac{3}{8}-\frac{27}{8}t^2+\frac{21}{2}t^3-\frac{15}{2}t^4, && t\in[0,1]. \end{aligned}$$

The three functions are chosen in such a way that:

- $x_1^\dagger$ violates the low-order reference source condition (3.11), i.e. $J_q(x_1^\dagger)\notin\mathcal{R}(A^\star)$,
- $x_2^\dagger$ fulfills the low-order reference source condition (3.11) but violates the high-order reference source condition (3.29), i.e. $J_q(x_2^\dagger)\in\mathcal{R}(A^\star)$ but $J_q(x_2^\dagger)\notin\mathcal{R}(A^\star J_p(A\,\cdot))$ whereas
- $x_3^\dagger$ fulfills the high-order reference source condition (3.29), i.e. $J_q(x_3^\dagger)\in\mathcal{R}(A^\star J_p(A\,\cdot))$.

In the Hilbert space situation $\mathcal{X}=\mathcal{Y}=L^2(0,1)$ we can further characterize the underlying source conditions. Following [44, Example 4] we have $x_1^\dagger\in\mathcal{R}((A^*A)^\nu)$ for all $\nu<\frac{1}{4}$. Using this property we can derive $x_2^\dagger\in\mathcal{R}((A^*A)^\mu)$ for all $\mu<\frac{3}{4}$. Of course, for $x_3^\dagger$ we get $x_3^\dagger\in\mathcal{R}(A^*A)$ which is the source condition where Tikhonov regularization saturates.

We can also characterize the function $x_1^\dagger$ more precisely in Banach spaces. Remembering the construction of the distance function for $x\equiv 1$ in the low-order case, see Example 3.2.1, the violation of the boundary conditions $x(1)=0$ was the crucial point in the derivation of the bound for the distance function. The same effect occurs here. So we can derive a similar bound for the distance function $d(R;J_q(x_1^\dagger))$, i.e. we can assume

$$d(R;J_q(x_1^\dagger))\le C\,R^{-\frac{q-1}{q}p}, \qquad R\ge 1,$$

for some constant $C>0$. An analogous consideration in the high-order case leads to the observation that we can suppose the distance function

$$d_p(R;J_q(x_1^\dagger))\le C\,R^{-\frac{q-1}{p(q-1)+1}}, \qquad R\ge 1,$$

with respect to the high-order reference source condition (3.29), see Example 3.3.4. This gives us a chance to compare theoretical and numerical results.

Additionally we use the same discretization as describes in the introducing Section 1.2. Here we choose $n=1000$ in the noiseless case and $n=500$ in the case of noisy data.

3.6.3 Numerical results I – noiseless data

We present some numerical results. We start with the sample function $x_1^\dagger$. In Figure 3.5 we see the error $\|x_\alpha-x^\dagger\|$ for $\alpha\in[10^{-5},10^{-2}]$ with choices $p=2$ and $q=1.1,1.5,2.0,2.5$ in the left and with $q=2$ and $p=1.1,1.5,2.0,2.5$ in the right picture.

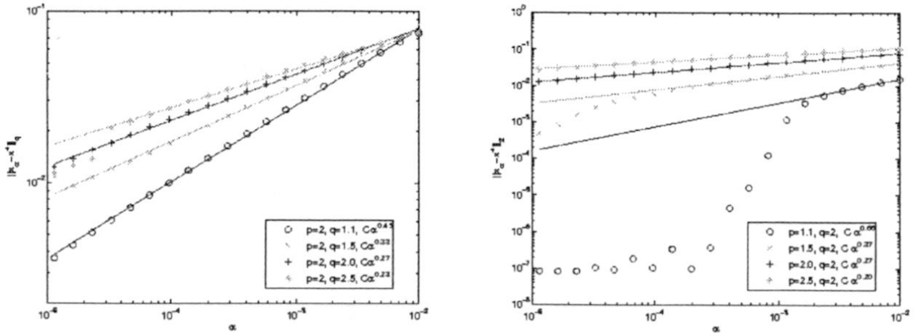

Figure 3.5: Convergence rates for given exact data, sample function $x_1^\dagger$ for varying parameter q (left) and varying parameter p (right)

	$p=2$			
	$q=1.1$	$q=1.5$	$q=2.0$	$q=2.5$
low-order	0.08	0.20	0.25	0.20
high-order	0.07	0.19	0.25	0.19
numeric.	0.45	0.33	0.27	0.23

	$q=2$			
	$p=1.1$	$p=1.5$	$p=2.0$	$p=2.5$
low-order	0.45	0.33	0.25	0.20
high-order	0.45	0.33	0.25	0.20
numeric.	0.66	0.37	0.27	0.20

Table 3.3: Theoretical convergence rates vs. practical results for $x_1^\dagger$ – noiseless data

In Table 3.3 we compare the theoretical convergence rates results of the previous section with the corresponding numerical results. We observe that the numerical convergence rates are in all cases somewhat better then theoretically predicted. In particular, for for varying parameter q we remark a relatively strong discrepancy between theoretical and practical results. How can we explain the difference? One key point might the following simple observation:

- Let $\tilde{\mathcal{X}}$ be another Banach space with $\tilde{\mathcal{X}} \cap \mathcal{X} \neq \emptyset$ and $\tilde{\mathcal{X}}^* \cap \mathcal{X}^* \neq \emptyset$ satisfying:
  - the duality products are equivalent on $\tilde{\mathcal{X}}^* \cap \mathcal{X}^* \times \tilde{\mathcal{X}} \cap \mathcal{X}$, i.e.

$$\langle x^*, x\rangle_{\tilde{\mathcal{X}}^*,\mathcal{X}^*} = \langle x^*, x\rangle_{\tilde{\mathcal{X}},\mathcal{X}} \quad \text{for all } x^* \in \tilde{\mathcal{X}}^* \cap \mathcal{X}^* \text{ and all } x \in \tilde{\mathcal{X}} \cap \mathcal{X},$$

  - we have $x^\dagger \in \tilde{\mathcal{X}} \cap \mathcal{X}$ with $\xi^\dagger = P'(x^\dagger) \in \tilde{\mathcal{X}}^*$,
  - it holds $x_\alpha \in \tilde{\mathcal{X}} \cap \mathcal{X}$ for all $\alpha \in (0, \alpha_{max}]$ (respectively $x_\alpha^\delta \in \tilde{\mathcal{X}}$ in the case of noisy data) and

- there exists a constant $C > 0$ such that $\|x_\alpha - x^\dagger\|_{\tilde{\mathcal{X}}} \le C\,\|x_\alpha - x^\dagger\|_{\mathcal{X}}$ holds for all $\alpha \in (0, \alpha_{max}]$.
- There exists a linear bounded operator $\tilde{A} : \tilde{\mathcal{X}} \longrightarrow \mathcal{Y}$ which satisfies $\tilde{A}\,x = A\,x$ for all $x \in \tilde{\mathcal{X}} \cap \mathcal{X}$.

Then the following holds: defining the distance function

$$\tilde{d}(R) = \tilde{d}(R;\xi^\dagger) := \inf\left\{\|\xi^\dagger - \tilde{A}^*\omega\|_{\tilde{\mathcal{X}}^*} \;:\; \|\omega\| \le R\right\}, \qquad R \ge 0,$$

we can replace the original distance function $d(R)$ in the analysis of the previous sections by $\tilde{d}(R)$. We only have to replace the duality product $\langle \xi^\dagger, x^\dagger - x_\alpha\rangle_{\mathcal{X}^*,\mathcal{X}}$ by the equivalent term $\langle \xi^\dagger, x^\dagger - x_\alpha\rangle_{\tilde{\mathcal{X}}^*,\tilde{\mathcal{X}}}$. The consequence is quite clear: assume all conditions stated above are fulfilled. If then the distance functions $\tilde{d}(R)$ decays faster to zero than $d(R)$ as $R \to \infty$ then the practical convergence rates becomes better than in the previous sections proposed.

This is exactly what seems to happen here. In our situation (for fixed space $\mathcal{Y} = L^2(0,1)$) we can choose $\tilde{X} = L^2(0,1)$. Since all function $x^\dagger, x_\alpha$ are bounded in the L^∞-norm (at least in the discretized situation) we only have to observe that the estimate $\|x_\alpha - x^\dagger\|_{\tilde{\mathcal{X}}} \le C\,\|x_\alpha - x^\dagger\|_{\mathcal{X}}$ holds for all $\alpha > 0$. Such estimate always holds for $q > 2$ by the known embeddings. Of course, such estimate cannot be true for arbitrary $x \in L^2(0,1) \cap L^q(0,1)$ for $q < 2$. However, at least in the discretized setting we can apply the norm equivalence of finite dimensional spaces to derive the corresponding estimate. So it is left open if this observed effect is based only on discretization or also occur in the infinite dimensional situation.

Setting $\tilde{X} = L^2(0,1)$ we derive $\tilde{d}(R) \le \mathcal{C}\,R^{-\frac{p}{2}}$. Repeating the calculations of the previous section for the low-order situation we derive theoretical convergence rates $\|x_\alpha - x^\dagger\|_{\mathcal{X}} \le \mathcal{C}\,\alpha^{\tilde{\nu}_\alpha}$ and $\|x_\alpha^\delta - x^\dagger\|_{\mathcal{X}} \le \mathcal{C}\,\delta^{\tilde{\nu}_\delta}$ with exponents $\tilde{\nu}_\alpha > 0$ and $\tilde{\nu}_\delta > 0$ respectively which are presented in Table 3.4. Table 3.5 now compares these theoretical considerations with

	$q \le 2$	$q \ge 2$
$\tilde{\nu}_\alpha$	$\frac{1}{2p}$	$\frac{1}{q(p-1)+2(p-1)}$
$\tilde{\nu}_\delta$	$\frac{p}{2(p+1)}$	$\frac{p}{q(p+2)-2}$

Table 3.4: Theoretical convergence rates for $x_1^\dagger$ in the low-order case, alternative distance function $\tilde{d}(R)$

practical results. Here, theoretical and numerical convergence rates $\tilde{\nu}_\alpha$ for the norm-convergence as well as the rates $\max\{2,q\}\tilde{\nu}_\alpha$ for convergence with respect to the Bregman distance $D_{\xi^\dagger}(x_\alpha, x^\dagger)$ are presented. We observe the following remarkable effect: whereas the numerical norm-convergence rates still differ from the theoretical results we contrariwise notice that the numerical and theoretical rates with respect to the Bregman distances coincide rather well. Since the convergence rates analysis essentially bases on estimates of Bregman distances the discrepancy of theory and numerics with respect to the norm-convergence might have the following reason: the exponent $s := \max\{2, q\}$ in the estimate $D_{\xi^\dagger}(x_\alpha, x^\dagger) \geq C \|x_\alpha - x^\dagger\|_{L^q}^s$ coming from the s-convexity of the space $L^p(0,1)$ can be improved in this specific situation. The convergence rates with respect to the Bregman distances are presented in Figure 3.6. Here additionally the behavior of the L^2-estimates $\|x_\alpha - x^\dagger\|_{L^2}$ are plotted which shows that the numerical convergence rates with respect to the L^2-norm do not depend on the choice of the parameter q. We also want to emphasize

		$p = 2$			
		$q = 1.1$	$q = 1.5$	$q = 2.0$	$q = 2.5$
$\tilde{\nu}_\alpha$	low-order	0.25	0.25	0.25	0.20
	numeric.	0.45	0.33	0.27	0.23
$\max\{2, q\}\tilde{\nu}_\alpha$	low-order	0.5	0.5	0.5	0.5
	numeric.	0.50	0.49	0.54	0.53

Table 3.5: Theoretical convergence rates vs. practical results for $x_1^\dagger$ – noiseless data, alternative distance function $\tilde{d}(R)$

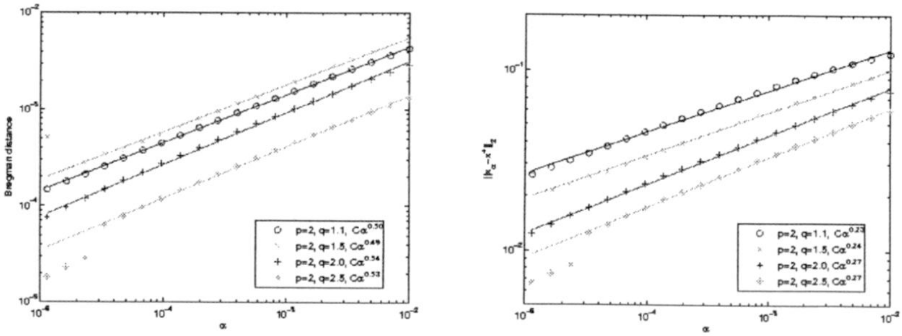

Figure 3.6: Convergence rates for given exact data with respect to the Bregman distance (left) and the L^2-norm (right), sample function $x_1^\dagger$ for varying parameter q

the case $p = 1.1$ and $q = 2$. Here, the convergence rate nearly behaves like the case $p = 1$ and fulfilled source condition (3.11): if α is chosen sufficiently small, i.e. $\alpha \leq \|\omega\|$ where $J_q(x^\dagger) = A^\star\omega$, $\omega \in \mathcal{Y}^*$, then $x_\alpha = x^\dagger$ holds. We remark that by bijectivity of the discretized problem the discretized version of source condition (3.11) is always satisfied.

In the following the results for the functions $x_2^\dagger$ and $x_3^\dagger$ are plotted. We again observe that for $p = 1.1$ the regularization error $\|x_\alpha - x_j^\dagger\|$, $j = 2, 3$, vanishes up to a numerical error. We can compare theoretical and numerical results for the sample function $x_3^\dagger$. Since the high-order reference source condition (3.29) is fulfilled we can apply the results

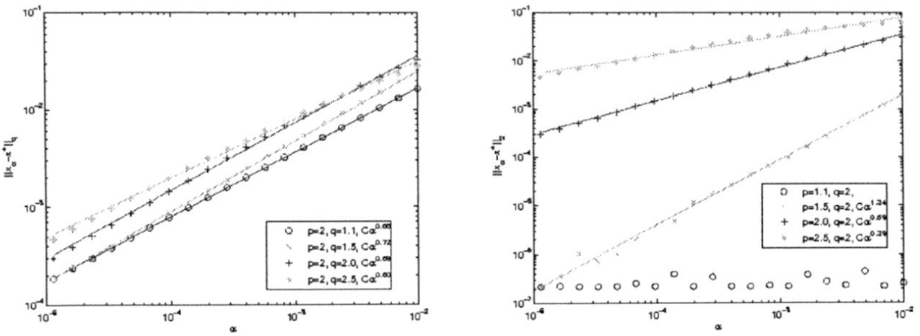

Figure 3.7: Convergence rates for given exact data, sample function $x_2^\dagger$ for varying parameter q (left) and varying parameter p (right)

	$p = 2$			
	$q = 1.1$	$q = 1.5$	$q = 2.0$	$q = 2.5$
theor.	0.55	0.75	1.0	0.80
numeric.	0.65	0.68	0.88	0.75

	$q = 2$			
	$p = 1.1$	$p = 1.5$	$p = 2.0$	$p = 2.5$
theor.	10	2.00	1.00	0.67
numeric.	–	1.16	0.88	0.52

Table 3.6: Theoretical convergence rates vs. practical results for $x_3^\dagger$ – noiseless data

of Lemma 3.3.9 with $\delta = 0$ and $d = 0$. Then we immediately observe a convergence rate $\|x_\alpha - x^\dagger\| \leq C\,\alpha^{\frac{\min\{2,q\}}{\max\{2,q\}(p-1)}}$ for some constant $C > 0$. We present numerical and theoretical results in Table 3.6. We see that the numerical results in general are somewhat worse than predicted. Only for the situation $p = 2$, $q = 1.1$ we achieve a better numerical convergence rate. But this might be explained with similar arguments with respect to the s-convexity as for the sample function $x_1^\dagger$.

3.6.4 Numerical results II – noisy data

We examine the case of noisy data. As opposite to the notation in the theoretical part here $\delta = \delta_{rel}$ describes a relative perturbation of the exact data y. For the size of the perturbation we have chosen the interval $[10^{-4}, 10^{-2}]$. In the numerical study we perturb the exact data with a (fixed) random vector: let e be a piece-wise linear functions with

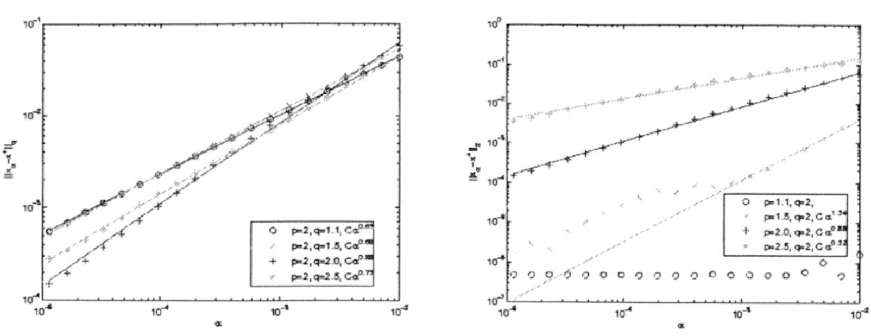

Figure 3.8: Convergence rates for given exact data, sample function $x_3^\dagger$ for varying parameter q (left) and varying parameter p (right)

Gaussian variables $e(j/n) \sim N(0,1)$, $1 \le j \le n = 500$, and set

$$y^\delta := y + \frac{e}{\|e\|}\|y\|\delta_{rel}, \quad \delta_{rel} \ge 0.$$

Again, as in the introducing example we calculate regularized solutions x_α^δ for a number of regularization parameters $\alpha_1 > \alpha_2 > \ldots > \alpha_l > 0$, $l \in \mathbb{N}$, and choose $\alpha = \alpha_{op}$ such that

$$\|x_\alpha^\delta - x^\dagger\| = \min\left\{\|x_{\alpha_j}^\delta - x^\dagger\| \;:\; 1 \le j \le l\right\}$$

for given exact solution $x^\dagger \in \mathcal{X}$. The numerical results are presented below.

We start again with sample function $x_1^\dagger$. The achieve rates are plotted in Figure 3.9. Additionally, we compare theoretical and numerical rates in Table 3.7. For calculating the numerical rates values only in such intervals were taken into account which feature a realistic behavior of the error $\|x_\alpha^\delta - x_j^\dagger\|$, $j = 1, 2, 3$, with respect to the noise-level δ. For noisy data additional numerical effects occurs which are a consequence of the worse attainability (or even unattainability) of the noisy data y^δ. So the Tikhonov functionals has worse mathematical properties which leads to a growing influence of numerical effects such as discretization errors and stopping criterions for the numerical algorithms on the corresponding solution x_α^δ. We observe in particular for $q = 2$ and varying parameter p a gap between theory and numerics.

Finally, the results for the functions $x_2^\dagger$ and $x_3^\dagger$ are plotted in Figure 3.10 and Figure 3.11. We see, that in particular in the cases $p = 2$, $q = 2.5$ and $p = 2.5$, $q = 2$ some numerical difficulties occur. Regarding the case $p = 2.5$, $q = 2$ for the sample function $x_2^\dagger$ no explainable dependency of the error $\|x_\alpha^\delta - x_2^\dagger\|$ on the noise-level δ can be detected. On the other hand we we at least that the rates of convergence in trend grow with respect to the smoothness of the underlying exact solution even the theoretical rates were not achieved for the smoothest sample function $x_3^\dagger$.

	$p=2$			
	$q=1.1$	$q=1.5$	$q=2.0$	$q=2.5$
theor.	0.33	0.33	0.33	0.25
numeric.	0.36	0.31	0.33	0.34

	$q=2$			
	$p=1.1$	$p=1.5$	$p=2.0$	$p=2.5$
theor.	0.26	0.30	0.33	0.35
numeric.	0.14	0.23	0.33	0.27

Table 3.7: Theoretical convergence rates vs. practical results for $x_1^\dagger$ – noisy data, alternative distance function

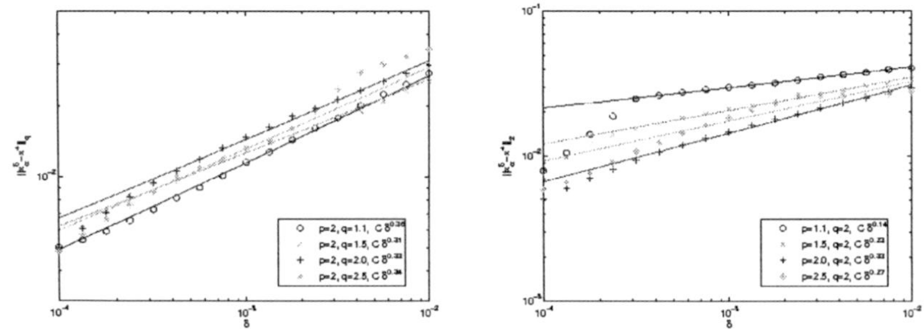

Figure 3.9: Convergence rates for given noisy data, sample function $x_1^\dagger$ for varying parameter q (left) and varying parameter p (right)

Again, for $x_3^\dagger$ we again apply the validity of the high-order reference source condition (3.29). Then we can use Theorem 3.3.1 to obtain the convergence rate $\|x_\alpha^\delta - x^\dagger\| \le C\,\delta^{\nu_\delta}$ with

$$\nu_\delta := \frac{\min\{2,p\}\min\{2,q\}}{\max\{2,q\}\,(\min\{2,q\}+\min\{2,p\}+1)}.$$

The accordant results are presented in Table 3.8. As we can observe the numerical results are somewhat worse than theoretically predicted also in this situation.

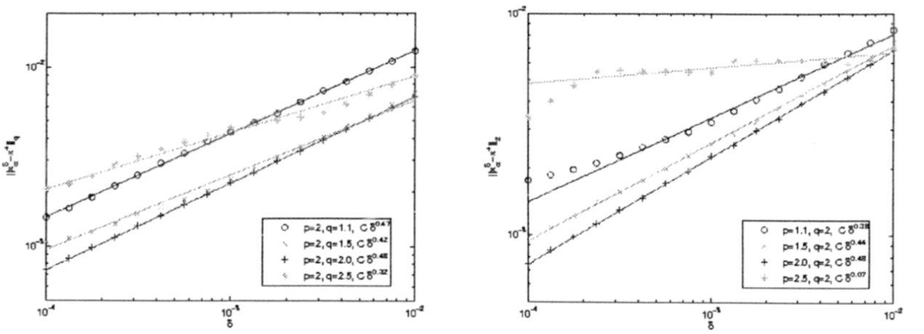

Figure 3.10: Convergence rates for given noisy data, sample function $x_2^\dagger$ for varying parameter q (left) and varying parameter p (right)

	$p = 2$			
	$q = 1.1$	$q = 1.5$	$q = 2.0$	$q = 2.5$
theor.	0.53	0.60	0.67	0.46
numeric.	0.47	0.47	0.57	0.34

	$q = 2$			
	$p = 1.1$	$p = 1.5$	$p = 2.0$	$p = 2.5$
theor.	0.53	0.60	0.67	0.67
numeric.	0.34	0.46	0.57	0.53

Table 3.8: Theoretical convergence rates vs. practical results for $x_3^\dagger$ – noisy data

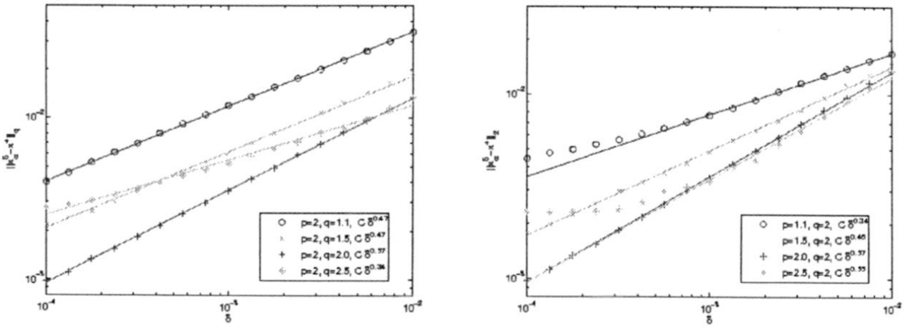

Figure 3.11: Convergence rates for given noisy data, sample function $x_3^\dagger$ for varying parameter q (left) and varying parameter p (right)

Chapter 4

Iterative regularization methods

4.1 Introduction

Major drawback of Tikhonov regularization is the high numerical effort. For the proper determination of the regularization parameter α we have to solve usually several non-quadratic minimization problems (exactly). Therefore the development and analysis of iterative regularization methods in Banach spaces are of high interest.

The theory of iterative regularization methods in Hilbert spaces has been deeply studied in the recent years. For a short overview we refer to [24], for more detailed information to [3] and [53]. Most of these algorithms can be described as a stable iteration procedure

$$x_{n+1}^\delta := x_n^\delta + B_n(x_n^\delta, y^\delta), \qquad n \in \mathbb{N},$$

in combination with an appropriate stopping criterion. Here, $B_n : \mathcal{X} \times \mathcal{Y} \longrightarrow \mathcal{X}$, $n \in \mathbb{N}$, denote in general nonlinear operators which are continuous with respect to both arguments. In particular, two different classes of iterative methods have been well-established:

- *Gradient-type methods* are regarded as easy to implement but providing only slow convergence. Here the iteration process is given by

$$B_n(x_n^\delta, y^\delta) := -\gamma_n F'(x_n^\delta)^* \left(F(x_n^\delta) - y^\delta\right).$$

The choice $\gamma_n \equiv 1$ is known as (nonlinear) *Landweber iteration* which has been studied in [62] and [21, Section 6.1] for linear and in [28] for nonlinear inverse problems. Accelerated versions such as the methods of *steepest descent* and *minimal error* were examined in [77] and [85]. A further modified version, i.e.

$$B_n(x_n^\delta, y^\delta) := -F'(x_n^\delta)^* \left(F(x_n^\delta) - y^\delta\right) - \beta_n(x_n^\delta - x_0).$$

for given sequence $\{\beta_n\}$ and initial guess $x_0 = x_0^\delta$ was considered in [86]. In [77] also a two-step algorithm can be found where the operators B_n additionally depend on x_{n-1}^δ.

- On the other hand, *Newton-type methods* promises faster convergence even the numerical effort in each iteration is higher. Let $\{g_\alpha\}$, $\alpha > 0$, denote any (linear) regularization as defined in Section 2.4. Then the choices

$$B_n(x_n^\delta, y^\delta) := g_{\alpha_n}(F'(x_n^\delta)^* F'(x_n^\delta)) F'(x_n^\delta)^* \left(y^\delta - F(x_n^\delta)\right)$$

are known as *inexact Newton-methods*, see [83] and [84]. Here $\{\alpha_n\}$ denotes a sequence of regularization parameters which can be chosen either a-priori or a-posteriori in each iteration step after some given rule. Most prominent example is the *Levenberg-Marquardt method* which was considered in [27] in the context of regularization theory. On the other hand the choices

$$B_n(x_n^\delta, y^\delta) := x_0 - x_n^\delta + g_{\alpha_n}\left(F'(x_n^\delta)^* F'(x_n^\delta)\right) F'(x_n^\delta)^* \left(y^\delta - F(x_n^\delta) - F'(x_n^\delta)(x_0 - x_n^\delta)\right)$$

are regarded to as *regularized Newton methods*, see [51]. Choosing Tikhonov regularization as regularization method we also can write

$$B_n(x_n^\delta, y^\delta) := \left(F'(x_n^\delta)^* F'(x_n^\delta) + \alpha_n I\right)^{-1} \left(F'(x_n^\delta)^* \left(y^\delta - F(x_n^\delta)\right) - \alpha_n(x_0 - x_n^\delta)\right).$$

This variant is called the *iteratively regularized Gauss-Newton method* which was originally introduced in [5] and further investigated e.g. in [8].

Here the focus is on gradient-type methods. We also refer to [54] for an iteratively regularized Gauss-Newton-type approach in Banach spaces. For given parameter $p > 1$ (with conjugate exponent $p^* > 1$) we reformulate equation (3.2) as minimization problem

$$\Omega_p(x) := \frac{1}{p}\|F(x) - y^\delta\|^p \to \min \quad \text{subject to} \quad x \in \mathcal{D}(F). \tag{4.1}$$

We generalize the results of [28] and [88] in a first step. Using a gradient method for solving the problem (4.1) we therefore deal with the following iteration:

$$\begin{aligned} x_0^\delta &:= x_0 = J_{s^*}^*(x_0^*) \in \mathcal{D}(F) \text{ with } x_0^* \in \mathcal{X}^*, \\ x_{n+1}^* &:= x_n^* - \mu_n F'(x_n^\delta)^* J_p\left(F(x_n^\delta) - y^\delta\right), \\ x_{n+1}^\delta &:= J_{s^*}^*(x_{n+1}^*), \end{aligned}$$

together with a proper choice of the step size μ_n and an appropriate stopping criterion. The choice of the parameter $s^* \in (1, 2]$ is determined by the supposed smoothness of the dual space $\mathcal{X}^*$. Moreover, $J_p : \mathcal{Y} \longrightarrow \mathcal{Y}^*$ and $J_{s^*}^* : \mathcal{X}^* \longrightarrow \mathcal{X}$ denote corresponding duality mappings with gauge functions $t \mapsto t^{p-1}$ and $t \mapsto t^{s^*-1}$ respectively. The algorithm above was considered in Banach spaces for linear operators in [88] and generalized to nonlinear problems in [54]. There, similar nonlinearity restrictions to operator F were applied as already supposed in [28] in the Hilbert space setting. We present here an analysis which is closely related to the one in [54] even the results are somewhat different. If $\mathcal{X}$ and $\mathcal{Y}$ are Hilbert spaces then the choices $p = 2$ and $\mu_n \equiv 1$ reduce the algorithm to classical Landweber iteration for nonlinear ill-posed problems which was originally considered in [28]. However, a constant step size leads usually to a slow convergence of gradient methods.

Therefore an appropriate choice of the parameter μ_n in each iteration step is crucial for a satisfying speed of convergence of the iteration process. We also point out that the update of the iterates (i.e. the search of the optimal step size μ_n) in fact takes place in the dual space $\mathcal{X}^*$.

We want to mention that the results in [88] were presented for uniform convex spaces $\mathcal{X}$ which is in fact a weaker condition than the p-convexity which is supposed here. On the other hand, uniform convex spaces such as L^r-spaces, $1 < r < \infty$, in practice are often also convex of power-type. Using this power-type convexity in the convergence analysis of the algorithm directly we can improve the choice of the parameter μ_n leading to faster convergence.

In a second step we choose $p = s$ (and write always p) and deal with algorithms of the form

$$\begin{aligned} x_0^\delta &:= x_0 = J_{p^*}^*(x_0^*) \quad \text{with} \quad x_0^* \in \mathcal{X}^*, \\ x_{n+1}^* &:= x_n^* - \mu_n F'(x_n^\delta)^* J_p\left(F(x_n^\delta) - y^\delta)\right) - \beta_n\left(x_n^* - x_\sharp^*\right), \\ x_{n+1}^\delta &:= J_{p^*}^*(x_{n+1}^*), \end{aligned}$$

where $x_\sharp^* \in \mathcal{X}^*$ denotes any a-priori guess. We should emphasize that here the convexity of the space $\mathcal{X}$ determines the choice of the exponent in (4.1) even the norm is the one in the space $\mathcal{Y}$. In Hilbert spaces this method is also known as regularized Landweber iteration which was introduced and investigated in [86]. However, as opposite to the approach therein we suggest here an a-posteriori choice of the parameter β_n and the step size μ_n. We will show that for properly chosen parameters $\{\mu_n\}$ and $\{\beta_n\}$ and under assumptions on the space $\mathcal{X}$ similar to those of [88] not only strong convergence but also convergence rates can be obtained. Moreover, we can give an extension of the convergence rates results of [86] in Hilbert spaces.

4.2 Preliminary assumptions

For the convergence analysis we need the following assumptions:

(B1) The Banach space $\mathcal{X}$ is supposed to be smooth and s-convex for some $s \in [2, \infty)$ and $\mathcal{Y}$ is assumed to be smooth.

(B2) For $\delta = 0$ there exists a solution $x^\dagger \in \mathcal{D}(F)$ of (3.2), i.e. $F(x^\dagger) = y$ holds. Moreover, the domain $\mathcal{D}(F)$ is supposed to be closed.

(B3) There exists a ball $\mathcal{B}_\varrho(x^\dagger)$ with radius $\varrho > 0$ around $x^\dagger$ such that for all $x \in \mathcal{B}_\varrho(x^\dagger) \cap \mathcal{D}(F)$ we can find a linear bounded operator $F'(x) : \mathcal{X} \longrightarrow \mathcal{Y}$ such that

   (i) For all $h \in \mathcal{X}$ satisfying $x + \varepsilon h \in \mathcal{D}(F)$ for all $0 \le \varepsilon \le \varepsilon_0$ for some $\varepsilon_0 = \varepsilon_0(h) > 0$ we have

$$F'(x)\, h := \lim_{\varepsilon \to 0} \frac{1}{\varepsilon}\left(F(x + \varepsilon\, h) - F(x)\right).$$

(ii) The operators $F'(x)$ are of degree $(1,0)$ of nonlinearity with uniform constant $0 \le L < 1$, i.e.

$$\|F(\tilde{x}) - F(x) - F'(x)(\tilde{x} - x)\| \le L\,\|F(\tilde{x}) - F(x)\| \tag{4.2}$$

holds for all $x, \tilde{x} \in \mathcal{B}_\varrho(x^\dagger) \cap \mathcal{D}(F)$.

(iii) It holds $\|F'(x)\| \le K$ uniformly for some constant $K > 0$ on $\mathcal{B}_\varrho(x^\dagger) \cap \mathcal{D}(F)$.

We shortly discuss these conditions. From the s-convexity of the (reflexive) space $\mathcal{X}$ we conclude the s^*-smoothness of the dual space $\mathcal{X}^*$. By the Xu/Roach inequalities (see Corollary 2.2.1) there exist constants $C_s > 0$ and $G^*_{s^*} > 0$ such that

$$\frac{1}{s}\|\tilde{x}\|^s - \frac{1}{s}\|x\|^s - \langle J_s(x), \tilde{x} - x\rangle \ge \frac{C_s}{s}\|x - \tilde{x}\|^s,$$

with duality mapping $J_s : \mathcal{X} \longrightarrow \mathcal{X}^*$ and

$$\frac{1}{s^*}\|\tilde{x}^*\|^{s^*} - \frac{1}{s^*}\|x^*\|^{s^*} - \langle J^*_{s^*}(x^*), \tilde{x}^* - x^*\rangle \le \frac{G^*_{s^*}}{s^*}\|x^* - \tilde{x}^*\|^{s^*}$$

hold for all $x, \tilde{x} \in \mathcal{X}$ and $x^*, \tilde{x}^* \in \mathcal{X}^*$ respectively. Both constants we will need in our convergence analysis. The smoothness of the space $\mathcal{Y}$ guarantees that duality mappings from $\mathcal{Y}$ into $\mathcal{Y}^*$ are always single valued.

Concerning assumption (B3) we remark the following. The nonlinearity restriction of type (4.2) with $L < \frac{1}{2}$ was already applied in [28] for dealing with Landweber iteration for nonlinear ill-posed problems in Hilbert spaces. We emphasize that in our convergence analysis the weaker condition $L < 1$ is sufficient. In particular, we make use of the inequality

$$\frac{1}{1+L}\|F'(x)(\tilde{x} - x)\| \le \|F(\tilde{x}) - F(x)\| \le \frac{1}{1-L}\|F'(x)(\tilde{x} - x)\|$$

for $x, \tilde{x} \in \mathcal{B}_\varrho(x^\dagger) \cap \mathcal{D}(F)$ which is an immediate consequence of (4.2). With $L = 0$ we also include the case of linear equations in our considerations. The introduced definition of the derivatives $F'(x)$ is again devoted the fact that in many application the assumption of interior points in the domain $\mathcal{D}(F)$ is too restrictive. On the other hand without the assumption $\mathcal{B}_\varrho(x^\dagger) \cap \mathcal{D}(F) \subset \mathcal{D}(F)$ it is hard to observe from the iteration process that all iterates remain in $\mathcal{D}(F)$. Therefore we pursue the following strategy: we show that all iterates remain in $\mathcal{B}_\varrho(x^\dagger)$ and assume additionally that these iterates belongs to $\mathcal{D}(F)$ (which is of course automatically satisfied if $x^\dagger \in \operatorname{int} \mathcal{D}(F)$ and ϱ sufficiently large).

4.3 Accelerated Landweber iteration

4.3.1 The algorithm

For $x^* \in \mathcal{X}^*$, $x = J^*_{s^*}(x^*)$ and arbitrary $\tilde{x} \in \mathcal{X}$ we keep the notation

$$\Delta_s(\tilde{x}, x) := \frac{1}{s}\|\tilde{x}\|^s - \frac{1}{s}\|x\|^s - \langle x^*, \tilde{x} - x\rangle,$$

for the Bregman distance of the functional $x \mapsto \frac{1}{s}\|x\|^s$.

We present the algorithm under consideration in detail:

Algorithm 4.3.1

(S0) Init. Choose start point $x_0^* \in \mathcal{X}^*$, $x_0^\delta := J_{s^*}^*(x_0^*)$ *with* $\Delta_s(x^\dagger, x_0^\delta) \leq \varrho^s \frac{C_s}{s}$. *Choose an upper bound* $\overline{\mu} \in (0, \infty]$ *for the step size and define the parameter* $\tau > 1$ *such that*

$$\tau > \frac{1+L}{1-L}$$

holds. Set $n := 0$.

(S1) STOP, if for $\delta > 0$ *the discrepancy criterion* $\|F(x_n^\delta) - y^\delta\| \leq \tau\,\delta$ *is fulfilled or we have* $F(x_n^\delta) = y$ *for* $\delta = 0$.

(S2) Calculate $\psi_n^* := F'(x_n^\delta)^\star J_p\left(F(x_n^\delta) - y^\delta\right)$,

$$\begin{aligned} C_{n,1} &:= \|F(x_n^\delta) - y^\delta\|^p(1-L) - (1+L)\delta\,\|F(x_n^\delta) - y^\delta\|^{p-1}, \\ C_{n,2} &:= \max\left\{G_{s^*}^*\|\psi_n^*\|^{s^*}, C_{n,1}\left(\overline{\mu}\,\|F(x_n^\delta) - y^\delta\|^{s-p}\right)^{-\frac{1}{s-1}}\right\} \quad \textit{and} \\ \mu_n &:= \left(\frac{C_{n,1}}{C_{n,2}}\right)^{s-1}. \end{aligned}$$

(S3) Calculate the new iterate

$$\begin{aligned} x_{n+1}^* &:= x_n^* - \mu_n \psi_n^* \\ x_{n+1} &:= J_{s^*}^*(x_{n+1}^*). \end{aligned}$$

Set $n := n+1$ *and go to step (S1).*

In the noiseless case we write x_n instead of $x_n^\delta = x_n^0$ for the iterates. Let $N(\delta, y^\delta)$ denote the index where the iteration stops. Then we have the relation

$$\left\|F(x_{N(\delta,y^\delta)}^\delta) - y^\delta\right\| \leq \tau\,\delta < \left\|F(x_n^\delta) - y^\delta\right\|, \quad 0 \leq n < N(\delta, y^\delta). \tag{4.3}$$

Furthermore, in particular the choice $\overline{\mu} = \infty$ is allowed in this algorithm. Then $C_{n,2} = G_{s^*}^*\|\psi_n^*\|^{s^*}$ holds automatically. We further introduce the constants

$$\underline{\mu}_\tau := \min\left\{\frac{(1-L-(1+L)\tau^{-1})^{s-1}}{G_{s^*}^{*\,s-1}K^s}, \overline{\mu}\right\} \quad \text{and} \quad \underline{\mu}_0 := \min\left\{\frac{(1-L)^{s-1}}{G_{s^*}^{*\,s-1}K^s}, \overline{\mu}\right\},$$

as well as

$$\lambda_\tau := \frac{1-L-(1+L)\tau^{-1}}{s} \quad \text{and} \quad \lambda_0 := \frac{(1-L)}{s}.$$

Then we can prove the following.

Lemma 4.3.1 *Assume (B1)-(B3) and all iterates* $\{x_n^\delta\}$ *remain in* $\mathcal{D}(F)$. *Then, for all* $n < N(\delta, y^\delta)$ *the following holds true:*

(i) The step size μ_n *is the (unique) maximizer of* $C(\mu) := C_{n,1}\mu - \frac{C_{n,2}}{s^*}\mu^{s^*}$, $\mu \in [0,\infty)$.

(ii) For $\delta > 0$ *we have* $\mu_n \in [\underline{\mu}_\tau, \overline{\mu}]\|F(x_n^\delta) - y^\delta\|^{s-p}$ *and*

$$C(\mu_n) \geq \lambda_\tau \mu_n \|F(x_n^\delta) - y^\delta\|^p \geq \lambda_\tau \underline{\mu}_\tau \|F(x_n^\delta) - y^\delta\|^s > 0.$$

(iii) For $\delta = 0$ *we have* $\mu_n \in [\underline{\mu}_0, \overline{\mu}]\|F(x_n) - y\|^{s-p}$ *and*

$$C(\mu_n) \geq \lambda_0 \mu_n \|F(x_n) - y\|^p \geq \lambda_0 \underline{\mu}_\tau \|F(x_n) - y\|^s > 0.$$

PROOF. The first part follows immediately by the definition of μ_n. Moreover, we observe

$$C(\mu_n) = \frac{C_{n,1}^s}{C_{n,2}^{s-1}} - \frac{1}{s^*}\frac{C_{n,1}^s}{C_{n,2}^{s-1}} = \frac{1}{s}\frac{C_{n,1}^s}{C_{n,2}^{s-1}} = \frac{1}{s}\mu_n C_{n,1}.$$

We now assume $\delta > 0$. Since the stopping criterion is not fulfilled we get $\tau\delta < \|F(x_n^\delta) - y^\delta\|$. Then we estimate

$$\begin{aligned} C_{n,1} &= \|F(x_n^\delta) - y^\delta\|^p \left[1 - L - (1+L)\frac{\delta}{\|F(x_n^\delta) - y^\delta\|}\right] \\ &\geq \|F(x_n^\delta) - y^\delta\|^p \left[1 - L - (1+L)\tau^{-1}\right] > 0. \end{aligned}$$

This automatically proves

$$C(\mu_n) = \frac{1}{s}\mu_n C_{n,1} \geq \lambda_\tau \mu_n \|F(x_n^\delta) - y^\delta\|^p.$$

Assume now $\overline{\mu} < \infty$ and $C_{n,2} = \left(\overline{\mu}\,\|F(x_n^\delta) - y^\delta\|^{s-p}\right)^{-\frac{1}{s-1}} C_{n,1}$. Then $\mu_n = \overline{\mu}\,\|F(x_n^\delta) - y^\delta\|^{s-p}$ holds by construction. Hence we have

$$C(\mu_n) = \frac{1}{s}\mu_n C_{n,1} \geq \|F(x_n^\delta) - y^\delta\|^p \frac{1 - L - (1+L)\tau^{-1}}{s}\mu_n = \lambda_\tau \overline{\mu} \|F(x_n^\delta) - y^\delta\|^s.$$

On the other side we now suppose $C_{n,2} = G_{s^*}^* \|\psi_n^*\|^{s^*}$. Since then it holds $C_{n,2} \geq \left(\overline{\mu}\,\|F(x_n^\delta) - y^\delta\|^{s-p}\right)^{-\frac{1}{s-1}} C_{n,1}$ we derive $\mu_n \leq \overline{\mu}\,\|F(x_n^\delta) - y^\delta\|^{s-p}$. With $\|F'(x_n^\delta)^\star\| = \|F'(x_n^\delta)\| \leq K$ we estimate

$$C_{n,2} \leq G_{s^*}^* \|F'(x_n^\delta)^\star\|^{s^*} \|J_p(F(x_n^\delta) - y^\delta)\|^{s^*} \leq G_{s^*}^* K^{s^*} \|F(x_n^\delta) - y^\delta\|^{s^*(p-1)}.$$

Hence we obtain

$$\begin{aligned} \mu_n = \left(\frac{C_{n,1}}{C_{n,2}}\right)^{s-1} &\geq \frac{\left(1 - L - (1+L)\tau^{-1}\right)^{s-1}}{G_{s^*}^{*\ s-1} K^s} \|F(x_n^\delta) - y^\delta\|^{(p - s^*(p-1))(s-1)} \\ &= \frac{\left(1 - L - (1+L)\tau^{-1}\right)^{s-1}}{G_{s^*}^{*\ s-1} K^s} \|F(x_n^\delta) - y^\delta\|^{s-p} \geq \underline{\mu}_\tau \|F(x_n^\delta) - y^\delta\|^{s-p} \end{aligned}$$

in that case. Consequently we can estimate

$$C(\mu_n) = \frac{1}{s}C_{n,1}\mu_n \geq \frac{1}{s}\|F(x_n^\delta) - y^\delta\|^s \left[1 - L - (1+L)\tau^{-1}\right] \underline{\mu}_\tau = \lambda_\tau \underline{\mu}_\tau \|F(x_n^\delta) - y^\delta\|^s.$$

This proves the second part. For $\delta = 0$ we have $C_{n,1} = \|F(x_n) - y\|^s(1 - L)$. Then an analogous calculation shows the third part of the assertions. ■

Remark 4.3.1 *Assume $F(x_n^\delta) \to y^\delta$ as $n \to \infty$. For $s > p$ we have $\|F(x_n^\delta) - y^\delta\|^{s-p} \to 0$ and hence $\mu_n \to 0$ as $n \to \infty$. On the other hand, for $s < p$ we conclude $\|F(x_n^\delta) - y^\delta\|^{s-p} \to \infty$ and consequently $\mu_n \to \infty$ as $n \to \infty$. However, in both cases we derive*

$$\|x_{n+1}^* - x_n^*\| = \mu_n \|\psi_n^*\| \le \overline{\mu}\, \|F(x_n^\delta) - y^\delta\|^{s-p} K\, \|F(x_n^\delta) - y^\delta\|^{p-1} = \overline{\mu} K\, \|F(x_n^\delta) - y^\delta\|^{s-1}$$

and hence $\|x_{n+1}^* - x_n^*\| \to 0$ as $n \to \infty$ as long as we have chosen $\overline{\mu} < \infty$.

We now show that as long as the discrepancy principle is not fulfilled the algorithm generates a decreasing sequence $\{\Delta_s(x^\dagger, x_n^\delta)\}$ of Bregman distances. Moreover, for $\delta > 0$ the algorithm terminates after a finite number $N(\delta, y^\delta)$ of iterations. We prove the following lemma.

Lemma 4.3.2 *Assume (B1)-(B3) and all iterates $\{x_n^\delta\}$ remain in $\mathcal{D}(F)$. Then, for all $0 \le n < N(\delta, y^\delta)$ the estimate*

$$\Delta_s(x^\dagger, x_{n+1}^\delta) < \Delta_s(x^\dagger, x_n^\delta)$$

is valid. In particular, $x_n^\delta \in \mathcal{B}_\varrho(x^\dagger)$ holds. Moreover, the following hold:

(i) If $\delta > 0$ then the algorithm stops after a finite number $N(\delta, y^\delta)$ of iterations. Moreover we have the estimate

$$N(\delta, y^\delta) \le C\, \delta^{-s}$$

for some constant $C > 0$ as well as

$$\sum_{n=0}^{N(\delta,y^\delta)-1} \mu_n \|F(x_n^\delta) - y^\delta\|^p \le \lambda_\tau^{-1} \Delta_s(x^\dagger, x_0)$$

and

$$\sum_{n=0}^{N(\delta,y^\delta)-1} \|F(x_n^\delta) - y^\delta\|^s \le \lambda_\tau^{-1} \underline{\mu}_\tau^{-1} \Delta_s(x^\dagger, x_0).$$

(ii) For $\delta = 0$ we assume that the algorithm does not stop after a finite number of iterations. Then we have

$$\sum_{n=0}^{\infty} \mu_n \|F(x_n) - y\|^p \le \lambda_0^{-1} \Delta_s(x^\dagger, x_0) \quad \text{and} \quad \sum_{n=0}^{\infty} \|F(x_n) - y\|^s \le \lambda_0^{-1} \underline{\mu}_0^{-1} \Delta_s(x^\dagger, x_0).$$

Proof. To shorten the notation we set $\Delta_n := \Delta_s(x^\dagger, x_n^\delta)$, $A_n := F'(x_n^\delta)$ and $A := F'(x^\dagger)$. By definition of the Bregman distance we have

$$\Delta_{n+1} := \frac{1}{s^*}\|x_{n+1}^*\|^{s^*} - \langle x_{n+1}^*, x^\dagger \rangle + \frac{1}{s}\|x^\dagger\|^s.$$

We continue with

$$\begin{aligned} \Delta_{n+1} - \Delta_n &= \frac{1}{s^*}\|x_{n+1}^*\|^{s^*} - \frac{1}{s^*}\|x_n^*\|^{s^*} - \langle x_{n+1}^* - x_n^*, x^\dagger \rangle \\ &= \frac{1}{s^*}\|x_{n+1}^*\|^{s^*} - \frac{1}{s^*}\|x_n^*\|^{s^*} - \langle -\mu_n A_n^* J_p(F(x_n^\delta) - y^\delta), x^\dagger \rangle. \end{aligned}$$

From the s^*-smoothness of $\mathcal{X}^*$ we conclude

$$\begin{aligned}\frac{1}{s^*}\|x^*_{n+1}\|^{s^*} - \frac{1}{s^*}\|x^*_n\|^{s^*} &\leq \langle x^\delta_n, -\mu_n A^*_n J_p(F(x^\delta_n) - y^\delta)\rangle + \frac{G^*_{s^*}}{s^*}\|\mu_n A^*_n J_p(F(x^\delta_n) - y^\delta)\|^{s^*} \\ &= \langle x^\delta_n, -\mu_n A^*_n J_p(F(x^\delta_n) - y^\delta)\rangle + \frac{G^*_{s^*}}{s^*}\mu_n^{s^*}\|\psi^*_n\|^{s^*}.\end{aligned}$$

Furthermore, we have with $T(x^\delta_n, x^\dagger) := F(x^\delta_n) - F(x^\dagger) - A_n(x^\delta_n - x^\dagger)$

$$\begin{aligned}&\langle x^\delta_n - x^\dagger, -\mu_n A^*_n J_p(F(x^\delta_n) - y^\delta)\rangle \\ &\quad = \mu_n\left(-\langle F(x^\delta_n) - y^\delta, J_p(F(x^\delta_n) - y^\delta)\rangle - \langle T(x^\delta_n, x^\dagger), J_p(F(x^\delta_n) - y^\delta)\rangle\right. \\ &\qquad \left.-\langle y^\delta - y, J_p(F(x^\delta_n) - y^\delta)\rangle\right) \\ &\quad \leq \mu_n\left(-\|F(x^\delta_n) - y^\delta\|^p + L\|F(x^\delta_n) - y\|\|F(x^\delta_n) - y^\delta\|^{p-1} + \delta\|F(x^\delta_n) - y^\delta\|^{p-1}\right) \\ &\quad \leq \mu_n\left(-(1-L)\|F(x^\delta_n) - y^\delta\|^p + (1+L)\delta\|F(x^\delta_n) - y^\delta\|^{p-1}\right)\end{aligned}$$

With the introduced notation we derived

$$\Delta_{n+1} \leq \Delta_n - \left(C_{n,1}\mu_n - \frac{C_{n,2}}{s^*}\mu_n^{s^*}\right).$$

Hence, from Lemma 4.3.1 we get

$$\Delta_{n+1} \leq \Delta_n - \lambda_0\mu_n\|F(x_n) - y\|^p \leq \Delta_n - \lambda_0\underline{\mu}_0\|F(x_n) - y\|^s$$

for $\delta = 0$ and

$$\Delta_{n+1} \leq \Delta_n - \lambda_\tau\mu_n\|F(x^\delta_n) - y^\delta\|^p \leq \Delta_n - \lambda_\tau\underline{\mu}_\tau\|F(x^\delta_n) - y^\delta\|^s$$

for $\delta > 0$. This proves the lemma by induction. We observe for $\delta > 0$ that

$$\Delta_0 \geq \Delta_0 - \Delta_{N(\delta,y^\delta)} \geq \lambda_\tau\underline{\mu}_\tau \sum_{n=0}^{N(\delta,y^\delta)-1} \|F(x^\delta_n) - y^\delta\|^s \geq \lambda_\tau\underline{\mu}_\tau N(\delta, y^\delta)(\tau\,\delta)^s$$

or equivalently $N(\delta, y^\delta) \leq \Delta_0\lambda_\tau^{-1}\underline{\mu}_\tau^{-1}\tau^{-s}\delta^{-s}$ which completes the proof. ■

Remark 4.3.2 *The use of Bregman distances here is opposite to their application in the analysis of Tihkonov functionals with $P(x) = \frac{1}{s}\|x\|^s$. Whereas in the last chapter we used the Bregman distances of form $\Delta_s(x, x^\dagger)$ we here deal with the Bregman distance $\Delta_s(x^\dagger, x)$ with interchanged arguments x and $x^\dagger$. Taking the non-symmetry of Bregman distances into account both approaches may vary.*

4.3.2 Convergence results

We now discuss the convergence properties of the algorithm. We start with the noiseless case $\delta = 0$. Here we can present the following convergence result.

Theorem 4.3.1 *Assume $\delta = 0$. Then the algorithm stops either after a finite number N of iterations with $\tilde{x}^\dagger := x_N$ or we have convergence $x_n \to \tilde{x}^\dagger$ as $n \to \infty$. In both cases the element $\tilde{x}^\dagger$ denotes any solution of equation (3.2) in $\mathcal{B}_\varrho(x^\dagger)$, i.e. we have $F(\tilde{x}^\dagger) = y$.*

PROOF. Let the iteration process do not stop after a finite number of steps. From Lemma 4.3.2 we conclude $F(x_n) \to y$ as $n \to \infty$. Hence we can choose indices $k > l$ such that $\|F(x_n) - y\| \geq \|F(x_k) - y\|$ for all $l \leq n \leq k$. Then we derive

$$\begin{aligned}
|\langle J_s(x_k) - J_s(x_l), x_k - x^\dagger\rangle| &= \left|\sum_{n=l}^{k-1}\langle x_{n+1}^* - x_n^*, x_k - x^\dagger\rangle\right| \\
&\leq \left|\sum_{n=l}^{k-1}\mu_n\langle J_p(F(x_n) - y), F'(x_n)(x_k - x_n + x_n - x^\dagger)\rangle\right| \\
&\leq \sum_{n=l}^{k-1}\mu_n\|F(x_n) - y\|^{p-1}(1+L)\left(\|F(x_n) - y\| + \|F(x_n) - F(x_k)\|\right) \\
&\leq \sum_{n=l}^{k-1}\mu_n\|F(x_n) - y\|^{p-1}(1+L)\left(2\|F(x_n) - y\| + \|F(x_k) - y\|\right) \\
&\leq 3\sum_{n=l}^{k-1}\mu_n\|F(x_n) - y\|^{p-1}(1+L)\|F(x_n) - y\| \\
&= 3\,(1+L)\sum_{n=l}^{k-1}\mu_n\|F(x_n) - y\|^p.
\end{aligned}$$

For $l \to \infty$ the right hand side goes to zero. Following the argumentation in [88] we have $\{x_n\}$ to be a Cauchy sequence and hence $x_n \to \tilde{x}^\dagger \in \mathcal{X}$. By the continuity of F and $F(x_n) \to y$ we have $F(\tilde{x}^\dagger) = y$ which shows that $\tilde{x}^\dagger$ is a solution. ■

Under an additional assumption we can characterize the limit element $\tilde{x}^\dagger \in \mathcal{D}(F)$. Therefore we cite the following result, see [54, Proposition 1].

Proposition 4.3.1 *Assume (B3). Then for all $x \in \mathcal{B}_\rho(x_0)$ with radius $\rho > 0$ chosen such that $\mathcal{B}_\rho(x_0) \subset \mathcal{B}_\varrho(x^\dagger)$ we have*

$$M_x := \{\tilde{x} \in \mathcal{B}_\rho(x_0) \;:\; F(\tilde{x}) = F(x)\} \subseteq (x + \mathcal{N}(F'(x))) \cap \mathcal{B}_\rho(x_0) \tag{4.4}$$

If additionally $\mathcal{B}_\rho(x_0) \subset \mathcal{D}(F)$ for some radius $\rho > 0$ then both sets are equal and

$$\mathcal{N}(F'(\tilde{x})) = \mathcal{N}(F'(x)), \qquad \forall\, \tilde{x} \in M_x, \tag{4.5}$$

holds.

The property (4.4) was already observed in [28, Proposition 2.1] in Hilbert spaces. Moreover, we need an additional condition which was introduced in [28] for the characterization of the limit element in a Hilbert space setting. There, the property

$$\mathcal{N}(F'(x^\dagger)) \subseteq \mathcal{N}(F'(x)) \qquad \text{for all} \quad x \in \mathcal{B}_\varrho(x^\dagger) \cap \mathcal{D}(F) \tag{4.6}$$

was supposed. Using this assumption we can prove the following result.

Theorem 4.3.2 *Assume (B1)-(B3) and $\mathcal{B}_\rho(x_0) \subset \mathcal{D}(F) \cap \mathcal{B}_\varrho(x^\dagger)$ for some radius $\rho > 0$. Then the following holds:*

(i) Assume the set $\{x \in \mathcal{D}(F) : F(x) = y\} \cap \mathcal{B}_\rho(x_0)$ to be non-empty. Then the minimization problem

$$\Delta_s(x, x_0) \to \min \quad \textit{subject to} \quad \{x \in \mathcal{D}(F) : F(x) = y\} \cap \mathcal{B}_\rho(x_0) \tag{4.7}$$

has a solution which is unique if this solution belongs to the interior of $\mathcal{B}_\rho(x_0)$.

(ii) Suppose $\delta = 0$. Assume $x^\dagger \in int\mathcal{B}_\rho(x_0)$ to be the (unique) solution of (4.7). Then $J_s(x^\dagger) - x_0^* \in \overline{\mathcal{R}(F'(x^\dagger)^\star)}$ holds. Moreover, if additionally (4.6) is valid, then we have convergence $x_n \to x^\dagger$, i.e. $x^\dagger = \tilde{x}^\dagger$ holds.

Proof. Assume $x \in \mathcal{D}(F)$ to be arbitrary chosen. From (4.4) we can conclude that $F(x) = y$ if and only if $x - x^\dagger \in \mathcal{N}(F'(x^\dagger)) = \mathcal{N}(F'(x))$. We examine the Bregman distance $\Delta_p(x, x_0)$. Let be $x, x_n \in \mathcal{X}$ arbitrary chosen with $x_n \rightharpoonup x$. By the weak lower semi-continuity of the norm we conclude

$$\frac{1}{s}\|x\|^s \le \liminf_{n\to\infty} \frac{1}{s}\|x_n\|^s.$$

Furthermore, $\langle x_0^*, x_n\rangle \to \langle x_0^*, x\rangle$ as $n \to \infty$ holds by definition of the weak convergence. Hence we have shown that

$$\Delta_s(x, x^\dagger) \le \liminf_{n\to\infty} \Delta_s(x_n, x_0),$$

i.e. the Bregman distance is sequentially weakly lower semi-continuous. Then the existence of a minimizer is clear by the coercivity of the Bregman distance $\Delta_s(x, x_0)$ and the weak closedness of the set $M := x^\dagger + \mathcal{N}(F'(x^\dagger)) \cap \mathcal{B}_\rho(x_0)$, see e.g. [98, Theorem 38.A and Corollary 38.14]. Let $x_\sharp$ denote a solution of (4.7). Considering the Bregman distance $\Delta_s(x, x_0)$ we have

$$\frac{\partial}{\partial x}\Delta_s(x, x_0) = \frac{\partial}{\partial x}\left(\frac{1}{s}\|x\|^s - \frac{1}{s}\|x_0\|^s - \langle x_0^*, x - x_0\rangle\right) = J_s(x) - x_0^* \in \mathcal{X}^*.$$

Assume $x_\sharp \in \mathcal{B}_\rho(x_0)$ and let $x \in \mathcal{N}(F'(x^\dagger))$ be chosen arbitrarily. Then $x_\sharp \pm \lambda x \in M$ for $\lambda > 0$ sufficiently small. From the optimality condition we conclude

$$0 \le \langle J_s(x_\sharp) - x_0^*, x_\sharp \pm \lambda x - x_\sharp\rangle = \pm\lambda\langle J_s(x_\sharp) - x_0^*, x\rangle$$

which implies

$$\langle J_s(x_\sharp) - x_0^*, x\rangle = 0, \qquad \forall x \in \mathcal{N}(F'(x^\dagger)).$$

Let $\mathcal{N}(F'(x))^\perp$ denote the annihilator of $\mathcal{N}(F'(x))$. Since $\overline{\mathcal{R}(F'(x^\dagger)^\star)} = \mathcal{N}(F'(x^\dagger))^\perp$ this proves $J_s(x_\sharp) - x_0^* \in \overline{\mathcal{R}(F'(x^\dagger)^\star)}$. Let $\tilde{x}_\sharp$ denote another solution of (4.7). Then

$$\langle J_s(x_\sharp) - J_s(\tilde{x}_\sharp), x\rangle = 0, \qquad \forall x \in \mathcal{N}(F'(x^\dagger)),$$

follows immediately. Setting $x := x_\sharp - \tilde{x}_\sharp$ this and the strict monotonicity of the duality mapping J_p, see Proposition 2.2.1(iii), imply $x_\sharp = \tilde{x}_\sharp$.

We consider the second part. We have to show $x^\dagger = \tilde{x}^\dagger$. From the theorem above we have $x_n \to \tilde{x}^\dagger$ and $x_n^* \to J_s(\tilde{x}^\dagger)$ as $n \to \infty$ with $F(\tilde{x}^\dagger) = y$. Since $\overline{\mathcal{R}(F'(x_n)^*)} = \mathcal{N}(F'(x_n))^\perp \subseteq \mathcal{N}(F'(x^\dagger))^\perp$, we see from the iteration process that $J_s(\tilde{x}^\dagger) - x_0^* \in \mathcal{N}(F'(x^\dagger))^\perp$. In particular, this implies

$$\left\langle J_s(\tilde{x}^\dagger) - x_0^*, x \right\rangle = 0, \qquad \forall\, x \in \mathcal{N}(F'(x^\dagger)).$$

Since $x^\dagger$ is the minimizer of (4.7) we immediately observe that

$$\langle J_s(x^\dagger) - J_s(\tilde{x}^\dagger), x \rangle = 0, \qquad \forall\, x \in \mathcal{N}(F'(x^\dagger)),$$

Setting $x := x^\dagger - \tilde{x}^\dagger \in \mathcal{N}(F'(x^\dagger))$, this again implies $x^\dagger = \tilde{x}^\dagger$. This proves the theorem. ■

If $x_0 \notin \operatorname{int} \mathcal{D}(F)$ we still can state the following.

Corollary 4.3.1 *Assume (B1)-(B3), $\delta = 0$ and $x_0 \notin \operatorname{int} \mathcal{D}(F)$. Moreover, suppose (4.5) and let $\tilde{x} \in \mathcal{D}(F)$ satisfying $F(\tilde{x}) = y$ be arbitrary chosen. Suppose further that the unique solution of the minimizing problem*

$$\Delta_s(x, x_0) \to \min \quad \textit{subject to} \quad \{x \in \mathcal{X} \,:\, x \in \tilde{x} + \mathcal{N}(F'(\tilde{x}))\}\,,$$

belongs to $\mathcal{D}(F)$. If $x^\dagger$ denotes this solution and (4.6) holds, then we have convergence $x_n \to x^\dagger$.

The proof is essentially the same as for Theorem 4.3.2.

Finally – under some additional assumptions – we present a stability result which shows that $x^\delta_{N(\delta,y^\delta)}$ depends stable on the given data.

Theorem 4.3.3 *Assume (B1)-(B3), $x^\dagger \in \operatorname{int} \mathcal{B}_\rho(x_0) \subset \mathcal{D}(F) \cap \mathcal{B}_\varrho(x^\dagger)$ for some radius $\rho > 0$. Suppose, furthermore, that F' depends continuously on x and $\mathcal{Y}$ is uniformly smooth. If (4.6) holds then we have convergence $x^\delta_{N(\delta,y^\delta)} \to x^\dagger$ as $\delta \to 0$.*

PROOF. Introducing a change in the notation we write $(x_n^\delta)^*$ for the iterates in the dual space $\mathcal{X}^*$ for noisy data and x_n^* for the case $\delta = 0$. Let n be a fixed index. Since $\mathcal{X}^*$ and $\mathcal{Y}$ are uniformly smooth the duality mappings $J^*_{s^*}$ and J_p are uniformly continuous on each bounded set, see e.g. [98, Proposition 47.19]. Hence, under the assumptions stated above the iterated x_n^δ and $(x_n^\delta)^*$ depend continuously on the given data y^δ. From Theorem 4.3.2(ii) we conclude $x_n \to x^\dagger$ as $n \to \infty$. Without loss of generality we can assume $N(\delta, y^\delta) \to \infty$ as $\delta \to 0$. Then, for $N(\delta, y^\delta) \geq n$ we obtain

$$\begin{aligned}
\Delta_s(x^\dagger, x^\delta_{N(\delta,y^\delta)}) &\leq \Delta_s(x^\dagger, x_n^\delta) \\
&= \frac{1}{s}\|x^\dagger\|^s - \frac{1}{s}\|x_n^\delta\|^s - \langle (x_n^\delta)^*, x^\dagger - x_n^\delta\rangle \\
&= \|x^\dagger\|^s - \frac{1}{s}\|x_n\|^s - \langle x_n^*, x^\dagger - x_n\rangle + \frac{1}{s}\|x_n\|^s - \frac{1}{s}\|x_n^\delta\|^s \\
&\qquad + \langle x_n^*, x^\dagger - x_n\rangle - \langle (x_n^\delta)^*, x^\dagger - x_n^\delta\rangle \\
&= \Delta_s(x^\dagger, x_n) + \frac{1}{s^*}\left(\|x_n^\delta\|^s - \|x_n\|^s\right) + \langle (x_n^\delta)^* - x_n^*, x^\dagger\rangle.
\end{aligned}$$

The right hand side vanishes for $\delta \to 0$ and $n \to \infty$. On the other hand, convergence $\Delta_s(x^\dagger, x^\delta_{N(\delta,y^\delta)}) \to 0$ as $\delta \to 0$ implies $x^\delta_{N(\delta,y^\delta)} \to x^\dagger$ as $\delta \to 0$ since $\mathcal{X}$ is supposed to be s-convex. This proves the theorem. ■

4.3.3 Further acceleration

The suggested choice of the step size parameter μ_n in Algorithm 4.3.1 has the advantage that it can be calculated explicitly. On the other hand, several (strong) estimates were necessary for deriving the underlying formulas. So we can expect to find alternative choices for the parameter $\mu = \mu_n$ leading to faster decay of the Bregman distances $\Delta_s(x^\dagger, x_n)$ and hence to a lower number of iterations $N(\delta, y^\delta)$.

Therefore we return to the minimization problem (4.1) with arbitrary parameter $p > 1$. Then we derive with $A_n^\star := F'(x_n^\delta)^\star$ and $\psi_n^* = A_n^\star J_p(F(x_n^\delta) - y^\delta)$

$$\begin{aligned}
\Delta_s(x^\dagger, J_{s^*}^*(x_n^* - \mu\psi_n^*)) - \Delta_n &= \frac{1}{s^*}\|x_n^* - \mu\,\psi_n^*\|^{s^*} - \frac{1}{s^*}\|x_n^*\|^{s^*} + \mu\,\langle A_n^\star J_p(F(x_n^\delta) - y^\delta), x^\dagger\rangle \\
&= \frac{1}{s^*}\|x_n^* - \mu\,\psi_n^*\|^{s^*} - \frac{1}{s^*}\|x_n^*\|^{s^*} + \mu\,\langle \psi_n^*, x_n^\delta\rangle \\
&\quad +\mu\,\langle J_p(F(x_n^\delta) - y^\delta), A_n^\star(x^\dagger - x_n^\delta)\rangle \\
&\le -\mu\,\left((1-L)\|F(x_n^\delta) - y^\delta\|^p - (1+L)\delta\|F(x_n^\delta) - y^\delta\|^{p-1}\right) \\
&\quad \frac{1}{s^*}\|x_n^* - \mu\,\psi_n^*\|^{s^*} - \frac{1}{s^*}\|x_n^*\|^{s^*} + \mu\,\langle \psi_n^*, x_n^\delta\rangle .
\end{aligned}$$

In the linear case even equality holds for given exact data, i.e. if we suppose $L = 0$ and $\delta = 0$. Moreover, this estimate seems to be sharper than the one in the proof of Lemma 4.3.2. Therefore we suggest the following choice of the step size μ_n: choose the parameter $\mu = \mu_n$ in such a way, that the right hand side of the above estimate becomes minimal (with respect to μ). We introduce the notations

$$c_n^\delta := (1-L)\|F(x_n^\delta) - y^\delta\|^p - (1+L)\delta\|F(x_n^\delta) - y^\delta\|^{p-1} \tag{4.8}$$

and

$$f(\mu) := \frac{1}{s^*}\|x_n^* - \mu\,\psi_n^*\|^{s^*} + \mu\,\langle \psi_n^*, x_n^\delta\rangle - \mu\, c_n^\delta .$$

Then we easily can prove the following lemma.

Lemma 4.3.3 *Assume (B1)-(B3) and $\psi_n^* \neq 0$. Then the minimization problem*

$$f(\mu) \to \min \qquad \textit{subject to} \quad \mu > 0 \tag{4.9}$$

has a unique solution μ^* as long as the discrepancy criterion is not fulfilled.

PROOF. Differentiating $f(\mu)$ we see

$$f'(\mu) = -\langle J_{s^*}^*(x_n^* - \mu\,\psi_n^*), \psi_n^*\rangle + \langle x_n^\delta, \psi_n^*\rangle - c_n^\delta .$$

By monotonicity of the duality mappings this function $f'(\mu)$ is strictly increasing. Since the stopping criterion is not fulfilled we have $c_n^\delta > 0$ which shows $f'(0) = -c_n^\delta < 0$. By continuity of $f'(\mu)$ there exists a unique element $\mu = \mu^* > 0$ satisfying the necessary optimality condition $f'(\mu) = 0$. ■

Therefore we now suggest the following modification of Algorithm 4.3.1.

Algorithm 4.3.2 *In Algorithm 4.3.1 we replace step (S2) by*

(S2') Calculate $\psi_n^* := F'(x_n^\delta)^* J_p(F(x_n^\delta) - y^\delta)$ *and find the solution* μ^* *of the equation*

$$f'(\mu) = 0, \qquad \mu \geq 0. \tag{4.10}$$

Set $\mu_n := \min\left\{\mu^*, \overline{\mu}\,\|F(x_n^\delta) - y^\delta\|^{s-p}\right\}$ *with* $\overline{\mu} \in (0, \infty]$.

Comparing the results of the previous subsections we can state the following consequence.

Corollary 4.3.2 *Assume (B1)-(B3) and all iterates* $\{x_n^\delta\}$ *remain in* $\mathcal{D}(F)$. *Then the following hold:*

(i) We set $\mu_n := \mu^*$ *where* μ_n *is the solution of the problem (4.9). Then all results of Lemma 4.3.2 remain true.*

(ii) We set $\mu_n := \min\{\mu^*, \overline{\mu}\,\|F(x_n^\delta) - y^\delta\|^{s-p}\}$, $0 < \overline{\mu} < \infty$. *Then all convergence and stability results of Theorem 4.3.1 and Theorem 4.3.3 remain true.*

PROOF. By Lemma 4.3.3 the parameter choices are well defined. Let $\tilde{\mu}_n$ denotes the parameter generated by Algorithm 4.3.1. Assume $\delta > 0$. Then we derive by definition of μ_n that

$$\begin{aligned}\Delta_{n+1} - \Delta_n \;\leq\; f(\mu^*) - \frac{1}{s^*}\|x_n^*\|^{s^*} \;&\leq\; f(\tilde{\mu}_n) - \frac{1}{s^*}\|x_n^*\|^{s^*} \\ &\leq\; -\left(C_{n,1}\tilde{\mu}_n - \frac{C_{n,2}}{s^*}\tilde{\mu}_n^{s^*}\right) \\ &\leq\; -\lambda_\tau \tilde{\mu}_n \|F(x_n^\delta) - y^\delta\|^p,\end{aligned}$$

which was the essential property for proving Lemma 4.3.2. For convergence and stability we additionally applied an upper bound $\overline{\mu}\|F(x_n^\delta) - y^\delta\|^{s-p}$ on the suggested choice of the step sizes μ_n and $\tilde{\mu}_n$. The case $\delta = 0$ follows similarly. ■

Remark 4.3.3 *We observe the following:*

- *Finding the parameter* μ^* *we have to solve the nonlinear equation (4.10) numerically. It cannot be calculated explicitly in general. Hence the price of a lower iteration number* $N(\delta, y^\delta)$ *is a higher numerical effort in each iteration step. So it might depend on the specific problem which algorithm is numerically more efficient.*
- *Since* $f'(\mu)$, $\mu \geq 0$, *is strictly increasing such algorithms for solving equation (4.10) are easy to implement. Either one uses a secant method or if the duality mapping* $J_{s^*}^*$ *is supposed to be differentiable we even can apply Newton's methods. This is e.g. the case for* $\mathcal{X} = L^r$ *with* $r \geq 2$.

4.3.4 Some numerical results

Based on two sample functions we want to compare the numerical effort of the algorithms described above by a small numerical experiment. Additionally we apply Landweber's method with constant step size $\mu_n \equiv const.$ in order to see the acceleration effect of the control of the step size. It turns out that the choice $\mu_n \equiv 1$ is too large in that situation. Therefore we have set $\mu_n \equiv 0.1$ which was motivated by the observation of the calculated step sizes μ_n of Algorithm 4.3.1.

We consider again the linear operator of Example 1.2.1. Furthermore we choose $\mathcal{X} := L^{1.1}(0,1)$ and $\mathcal{Y} := L^2(0,1)$. Moreover we set $p = 2$ which coincides with the power of convexity of $\mathcal{X}$. As discretization level we set $n := 1000$. We deal with the functions

$$x_1^\dagger(t) := 3\,(t-0.5)^2 + 0.2, \quad t \in (0,1), \qquad \text{and} \qquad x_2^\dagger(t) := \begin{cases} 5, & t \in [0.25, 0.27], \\ -3, & t \in [0.4, 0.45], \\ 4, & t \in [0.7, 0.73], \\ 0, & \text{else.} \end{cases}$$

Notice that $x_2^\dagger$ is the function of the introducing example in Section 1.2. For the discrepancy criterion we set $\tau := 1.2$ and $x_0 \equiv 0$ as initial guess. The number of iterations was limited by $n_{max} = 10^6$. Additionally we introduce $\overline{\mu} := 100$ as upper bound for the step size.

We now turn to the numerical results. The number of iterations as well as the calculation times are presented in Table 4.1 for $x_1^\dagger$ and in Table 4.2 for the second function $x_2^\dagger$. We summarize the results:

- Even not presented here the quality of the approximate solutions $x_{N(\delta,y^\delta)}$ does not depend on the specific algorithm. In all cases the achieved results were only somewhat worse than the Tikhonov-regularized solutions with penalty functional $P(x) := \frac{1}{q}\|x\|_{L^q}^q$, $q = 1.1$, and the regularization parameter α chosen in an optimal way (using the knowledge of $x_i^\dagger$, $i = 1, 2$).
- Choosing a constant step size $\mu_n \equiv const.$ the number $N(\delta, y^\delta)$ of necessary iterations grows rapidly when the noise level δ becomes smaller. For a relative noise level $\delta_{rel} = 10^{-4}$ the maximal number $n_{max} = 10^6$ of iterations was exceeded for both sample functions.
- Both accelerated versions lead to a strongly decrease of the iteration numbers. In particular, for moderate noise levels $\delta_{rel} = 10^{-3} \ldots 0.05$ (which are the one occurring in practical applications) the calculation times shows the good performance of the algorithms under consideration.
- For very small noise levels (or $\delta = 0$) the calculation times are still quite high. Here additional numerical stopping criterions should be tested leading to a earlier termination (which was not done here).
- The application of the forward operator A and its adjoint $A^\star$ was implemented here in an efficient way needing only $\mathcal{O}(n)$ operations. Using a matrix-vector multiplication for the implementation the differences between Algorithm 4.3.1 and 4.3.2 in time will increase since the calculation of the step size μ_n in Algorithm 4.3.2 uses only vector-vector operations (of order $\mathcal{O}(n)$).

Finally, we present a graphical demonstration of the effect of using Banach spaces in our considerations. Therefore we choose the Hilbert space $\mathcal{X} = L^2(0,1)$ in an alternative cal-

	$\mu_n = const.$		Algorithm 4.3.1		Algorithm 4.3.2	
δ_{rel}	$N(\delta, y^\delta)$	time (sec.)	$N(\delta, y^\delta)$	time (sec.)	$N(\delta, y^\delta)$	time (sec.)
0.05	863	0.85	63	0.16	28	0.18
0.01	7530	6.56	335	0.40	93	0.33
10^{-3}	79120	69.01	2065	2.29	451	2.05
10^{-4}	$> 10^6$	–	24548	26.27	2068	8.69
10^{-5}	–	–	118823	126.81	12479	49.97

Table 4.1: Calculation times for sample function $x_1^\dagger$

	$\mu_n = const.$		Algorithm 4.3.1		Algorithm 4.3.2	
δ_{rel}	$N(\delta, y^\delta)$	time (sec.)	$N(\delta, y^\delta)$	time (sec.)	$N(\delta, y^\delta)$	time (sec.)
0.05	4023	3.52	253	0.35	104	0.58
0.01	36720	31.98	1520	1.35	358	1.57
10^{-3}	457270	391.40	11022	12.01	963	4.17
10^{-4}	$> 10^6$	–	94315	101.77	6729	27.10
10^{-5}	–	–	606582	653.37	50890	205.01

Table 4.2: Calculation times for sample function $x_2^\dagger$

culation which leads back to classical (accelerated) Landweber iteration. The regularized solutions $x^\delta_{N(\delta,y^\delta)}$ for $x_2^\dagger$ and $\delta_{rel} = 0.01$ were plotted in Figure 4.1. We see that the choice $\mathcal{X} = L^q(0,1)$ has the same effect as the choice of the penalty functional $P(x) := \frac{1}{q}\|x\|^q_{L^q}$ with $q = 1.1$ or $q = 2$ in the Tikhonov functional. Consequently, the numerical effort as well as the obtained regularized solutions $x^\delta_{N(\delta,y^\delta)}$ of this numerical example show that accelerated Landweber methods are an interesting alternative to Tikhonov regularization with penalty terms based on Banach space norms.

4.4 Modified Landweber iteration

4.4.1 General remarks

We recall that we now choose $p = s$, i.e. the parameter p now also describes the power of convexity of the space $\mathcal{X}$. The presented convergence and convergence rates analysis is based essentially on the following variational inequality which is an immediate consequence of the concept of approximate source conditions. Based on the low order convergence rates analysis for Tikhonov regularization in Chapter 3.2 we introduce here the following reference source condition

$$\xi^\dagger - x_\sharp^* = F'(x^\dagger)^*\omega, \qquad \omega \in \mathcal{Y}^*, \tag{4.11}$$

for given $x_\sharp^* \in \mathcal{X}^*$ and $\xi^\dagger := J_p(x^\dagger)$. Then the following holds.

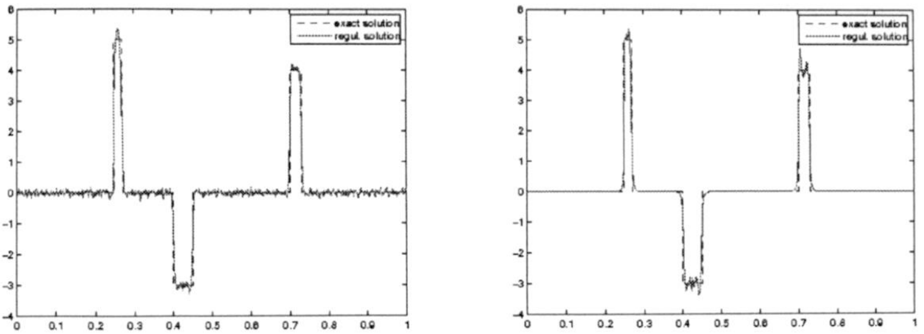

Figure 4.1: exact vs. regularized solution for $x_2^\dagger$ for $\mathcal{X} = L^2(0,1)$ (left plot) and $\mathcal{X} = L^{1.1}(0,1)$ (right plot)

Lemma 4.4.1 *Assume (B1)-(B3) and let the a-priori guess $x_\sharp^* \in \mathcal{X}^*$ be arbitrary chosen. Then there exists a decreasing function $d_0(R)$, $R \geq 0$, satisfying*

$$\left|\langle \xi^\dagger - x_\sharp^*, x - x^\dagger \rangle\right| \leq R(1+L)\left(\|F(x) - y^\delta\| + \delta\right) + \Delta_p(x^\dagger, x) + d_0(R)^{p^*} \tag{4.12}$$

for all $R \geq 0$ and all $x \in \mathcal{B}_\varrho(x^\dagger) \cap \mathcal{D}(F)$. The same estimate (4.12) holds if the Bregman distance $\Delta_p(x^\dagger, x)$ is replaced by $\Delta_p(x, x^\dagger)$. Moreover, if $\xi^\dagger - x_\sharp^* \in \overline{\mathcal{R}(F'(x^\dagger)^\star)}$ then we can choose $d_0(R) \to 0$ as $R \to \infty$.

PROOF. Let $d(R) := d(R; \xi^\dagger - x_\sharp^*)$, $R \geq 0$, denote the distance function of $\xi^\dagger - x_\sharp^*$ with recpect to the source condition (4.11). Following the lines of the proof of Lemma 3.2.1 with $\xi^\dagger - x_\sharp^*$ instead of $\xi^\dagger$, $c_2 = 0$ and $c_1 = \kappa = 1$ that

$$\begin{aligned}
\left|\langle \xi^\dagger - x_\sharp^*, x - x^\dagger \rangle\right| &\leq R\,(1+L)\,\|F(x) - y\| + d(R)\|x - x^\dagger\| \\
&\leq R\,(1+L)\left(\|F(x) - y^\delta\| + \delta\right) + d(R)\left(\frac{p}{C_p}\Delta_p(x^\dagger, x)\right)^{\frac{1}{p}} \\
&\leq R\,(1+L)\left(\|F(x) - y^\delta\| + \delta\right) + \Delta_p(x^\dagger, x) + \frac{1}{p^*}\left(\frac{1}{C_p}\right)^{\frac{1}{p-1}} d(R)^{p^*}
\end{aligned}$$

by applying Young's inequality in the last step. Hence inequality (4.12) holds whenever $d_0(R) \geq \left(\frac{1}{p^*}\right)^{\frac{1}{p^*}}\left(\frac{1}{C_p}\right)^{\frac{1}{p}} d(R)$. ■

In order to achieve a convergence rate result we additionally have to suppose that the distance function $d(R; \xi^\dagger - x_\sharp^*)$ tends to zero as $R \to \infty$, i.e.

$$\xi^\dagger - x_\sharp^* \in \overline{\mathcal{R}(F'(x^\dagger)^\star)} \tag{4.13}$$

holds. However, if we additionally assume condition (4.6), i.e. $\mathcal{N}(F'(x^\dagger)) \subseteq \mathcal{N}(F'(x))$ for all $x \in \mathcal{B}_\varrho(x^\dagger) \cap \mathcal{D}(F)$ and $x^\dagger$ is (uniquely) defined via

$$\Delta_p(x^\dagger, x_\sharp) := \min\left\{\Delta_p(x, x_\sharp) \;:\; F(x) = y\right\}$$

where $x_\sharp := J^*_{p^*}(x^*_\sharp) = J_p^{-1}(x^*_\sharp)$ then (4.13) holds, see Theorem 4.3.2.

As we will see, the proper choice of the function $d_0(R)$, $R \geq 0$, is crucial for the validity of the subsequent convergence and convergence rates results. In particular, the analysis of the second stopping criterion is based on the availability of the inequality

$$\frac{1}{p^*}\left(\frac{1}{C_p}\right)^{\frac{1}{p-1}} d(\bar{R};\xi^\dagger - x^*_\sharp)^{p^*} \leq d_0(\bar{R})^{p^*} \leq \bar{R}\,\delta \tag{4.14}$$

for some chosen parameter $\bar{R} = \bar{R}(\delta)$. Since for all $R \geq 0$ we have $d(R;\xi^\dagger - x^*_\sharp) \geq d(0;\xi^\dagger - x^*_\sharp) = \|\xi^\dagger - x^*_\sharp\|$ we can find a constant $C_{\xi^\dagger} > 0$ such that

$$C_{\xi^\dagger}\left(:= d_0(0)^{p^*}\right) \geq \frac{1}{p^*}\left(\frac{1}{C_p}\right)^{\frac{1}{p-1}} \|\xi^\dagger - x^*_\sharp\|^{p^*}. \tag{4.15}$$

Consequently, inequality (4.14) is satisfied automatically for all $\delta > 0$ by setting $\bar{R} := C_{\xi^\dagger}\delta^{-1}$. This gives us different ways for a proper choice of the parameter $\bar{R}$: if some a-priori information about $x^\dagger \in \mathcal{X}$ in form of the (distance) function $d_0(R)$, $R \geq 0$, is known we choose $\bar{R}$ according (4.14). For the case of non-availability of such information two possibilities arises: either one chooses $\bar{R} := C_{\xi^\dagger}\delta^{-1}$ which constant $C_{\xi^\dagger}$ chosen sufficiently large such that (4.15) holds or we set $\bar{R} := 0$ when the second inequality in (4.14) is not needed. In the latter case, inequality (4.12) is an immediate consequence of Young's inequality.

4.4.2 The general scheme

We first present the algorithm under consideration in this section in a general scheme. It is given as follows:

Algorithm 4.4.1

(S0) Init. Choose start point x^*_0, $x_0 := J^*_{p^*}(x^*_0)$, $D_0 > 0$, with $\Delta_p(x^\dagger, x_0) \leq D_0$, parameters $\tau_\beta > 0$, $\tau_\mu > 1$ and a decreasing function $d_0 : [0,\infty) \longrightarrow [0,\infty)$ such that (4.12) is fulfilled. Furthermore, choose $\bar{R} = \bar{R}(\delta) > 0$ either so big that $d_0(\bar{R})^{p^*} \leq \delta\,\bar{R} \leq d_0(0)^{p^*} =: C_{\xi^\dagger}$ or set $\bar{R} := 0$. Set $n := 0$.

(S1) Calculate $\psi^*_n := F'(x^\delta_n)^* J_p(F(x^\delta_n) - y^\delta)$ and $\phi^*_n := x^*_n - x^*_\sharp$. Compute μ_n, $C(\mu_n)$, γ_n,

$$\begin{aligned} \beta_n &:= \min\left\{1, \left(\frac{D_n}{\gamma_n}\right)^{p-1}\right\} \quad \textit{and} \\ D_{n+1} &:= (1-\beta_n)\,D_n - C(\mu_n) + \frac{\gamma_n}{p^*}\beta_n^{p^*}. \end{aligned}$$

STOP, if certain stopping criterion is fulfilled.

(S2) Update the iterates

$$\begin{aligned} x^*_{n+1} &:= x^*_n - \mu_n\psi^*_n - \beta_n\phi^*_n \quad \textit{and} \\ x^\delta_{n+1} &:= J^*_{p^*}(x^*_{n+1}). \end{aligned}$$

(S3) Set $n := n+1$ and go to (S1).

Let x_n^δ, ψ_n^* and ϕ_n^* be given in the current iteration step. For proving convergence results we introduce the notation

$$\Delta_{\mu,\beta} := \Delta_p(x^\dagger, x_n^* - \mu\psi_n^* - \beta\phi_n^*), \qquad \mu, \beta \geq 0.$$

Recalling the definition (4.8) of c_n^δ we present the following inequality which plays the central role in the subsequent analysis.

Lemma 4.4.2 *Assume (B1)-(B3) and inequality (4.12) holds. Then the inequality*

$$\begin{aligned} \Delta_{\mu,\beta} &\leq (1-\beta)\,\Delta_p(x^\dagger, x_n^\delta) - \mu c_n^\delta + 2^{p^*-1}\frac{G_{p^*}^*}{p^*}\left(\|\psi_n^*\|^{p^*}\mu^{p^*} + \|\phi_n^*\|^{p^*}\beta^{p^*}\right) \\ &\quad + \left(R(1+L)\left(\|F(x_n^\delta) - y^\delta\| + \delta\right) + d_0(R)^{p^*}\right)\beta \end{aligned} \tag{4.16}$$

holds for all $\mu, \beta \geq 0$, all $n > 0$ and all $R \geq 0$.

Proof. Again we use the notation $\Delta_n := \Delta_p(x^\dagger, x_n^\delta)$, $A_n := F'(x_n^\delta)$ and $A := F'(x^\dagger)$ of the last section and start with

$$\Delta_{\mu,\beta} := \frac{1}{p^*}\|x_n^* - \mu\psi_n^* - \beta\phi_n^*\|^{p^*} - \langle x_n^* - \mu\psi_n^* - \beta\phi_n^*, x^\dagger\rangle + \frac{1}{p}\|x^\dagger\|^p.$$

By the p^*-smoothness of $\mathcal{X}^*$ we derive

$$\frac{1}{p^*}\|x_n^* - \mu\psi_n^* - \beta\phi_n^*\|^{p^*} \leq \frac{1}{p^*}\|x_n^*\|^{p^*} + \langle x_n^\delta, -\mu\psi_n^* - \beta\phi_n^*\rangle + \frac{G_{p^*}^*}{p^*}\|\mu\phi_n^* + \beta\psi_n^*\|^{p^*}.$$

This gives

$$\Delta_{\mu,\beta} \leq \Delta_n + \langle x_n^\delta - x^\dagger, -\mu\psi_n^* - \beta\phi_n^*\rangle + \frac{G_{p^*}^*}{p^*}\|\mu\psi_n^* + \beta\phi_n^*\|^{p^*}.$$

From the proof of Lemma 4.3.2 we derive

$$\langle x_n^\delta - x^\dagger, -\mu\psi_n^*\rangle \leq -\mu c_n^\delta.$$

Moreover, we can rewrite

$$\langle x_n^\delta - x^\dagger, -\beta\phi_n^*\rangle = -\beta\langle x_n^\delta - x^\dagger, x_n^* - \xi^\dagger\rangle - \beta\langle x_n^\delta - x^\dagger, \xi^\dagger - x_\sharp^*\rangle.$$

The first term generates a negative term, since

$$-\beta\langle x_n^\delta - x^\dagger, x_n^* - \xi^\dagger\rangle = -\beta\Delta_n - \beta\Delta_p(x_n^\delta, x^\dagger).$$

For the second term we use the variational inequality (4.12) to get

$$\begin{aligned} -\beta\langle x_n^\delta - x^\dagger, \xi^\dagger - x_\sharp^*\rangle &\leq R\,(1+L)\,\|F(x_n^\delta) - y^\delta\|\beta + R\,(1+L)\,\delta\,\beta \\ &\quad + \Delta_p(x_n^\delta, x^\dagger)\,\beta + d_0(R)^{p^*}\beta. \end{aligned}$$

Additionally we can apply Jensen inequality to obtain

$$\|\mu\psi_n^* + \beta\phi_n^*\|^{p^*} \leq 2^{p^*-1}\left(\mu^{p^*}\|\phi_n^*\|^{p^*} + \beta^{p^*}\|\psi_n^*\|^{p^*}\right),$$

which proves the estimate (4.16). ■

We present the following general convergence result for fixed $\delta \geq 0$.

Theorem 4.4.1 *Suppose (B1)-(B3). Assume that as long as the iteration does not stop we can find constants* $C_{n,1}, C_{n,2} > 0$ *and* $0 < \gamma_n = \gamma_n(\mu) < \infty$ *such that*

$$\Delta_p(x^\dagger, x_{n+1}^\delta) \leq (1-\beta_n)\,\Delta_p(x^\dagger, x_n^\delta) - \left[C_{n,1}\mu - \frac{C_{n,2}}{p^*}\mu^{p^*}\right] + \frac{\gamma_n}{p^*}\beta_n^{p^*} \tag{4.17}$$

holds for all $\mu > 0$. *Then the choice*

$$\mu_n := \left(\frac{C_{n,1}}{C_{n,2}}\right)^{p-1} \quad \text{and} \quad C(\mu_n) := \frac{1}{p}\mu_n C_{n,1}$$

yields the following for all $n \geq 0$*:*

(i) We have

$$\sum_{k=0}^{n} C(\mu_k) \leq D_0.$$

In particular, we have $C(\mu_n) \to 0$ *as* $n \to \infty$.

(ii) If $\gamma_n(\mu_n) \leq \Gamma < \infty$ *holds uniformly then there exists a constant* $C > 0$ *such that*

$$\Delta_p(x^\dagger, x_n^\delta) \leq D_n \leq C\, n^{-\frac{1}{p-1}}.$$

In particular, if the iteration does not stop then we have convergence $x_n^\delta \to x^\dagger$ *as* $n \to \infty$.

(iii) If $0 < \gamma < \gamma_n(\mu_n) \leq \Gamma < \infty$ *holds uniformly then there exists a constant* $C_\beta > 0$ *such that*

$$\beta_n \leq C_\beta n^{-1}.$$

Proof. We set $\mu_n := (C_{n,1}/C_{n,2})^{p-1}$ which maximizes the term in the brackets of inequality (4.17). Then

$$C_{n,1}\mu_n - \frac{C_{n,2}}{p^*}\mu_n^{p^*} = \frac{C_{n,1}^p}{C_{n,2}^{p-1}}\left(1 - \frac{1}{p^*}\right) = \frac{1}{p}\mu_n C_{n,1} = C(\mu_n) > 0$$

holds. We have by assertion $\Delta_0 \leq D_0$ and assume $\Delta_n \leq D_n$. Then we have

$$\begin{aligned} \Delta_{n+1} &\leq (1-\beta_n)\,\Delta_n - C(\mu_n) + \frac{\gamma_n}{p^*}\beta_n^{p^*} \\ &\leq \max\{1-\beta_n, 0\}\, D_n - C(\mu_n) + \frac{\gamma_n}{p^*}\beta_n^{p^*} \end{aligned}$$

which proves $\Delta_{n+1} \leq D_{n+1}$. We remark that the choice of $\beta = \beta_n$ minimizes the term

$$\max\{1-\beta, 0\}\, D_n + \frac{\gamma_n}{p^*}\beta^{p^*} \quad \text{for} \quad \beta > 0.$$

Hence we get

$$D_{n+1} < (1-\beta_n)\, D_n + \frac{\gamma_n}{p^*}\beta_n^{p^*} \leq D_n.$$

From the definition of D_{n+1} we conclude

$$D_{n+1} := (1-\beta_n)\,D_n - C(\mu_n) + \frac{\gamma_n}{p^*}\beta_n^{p^*} \leq D_n - C(\mu_n)$$

or equivalently $C(\mu_n) \leq D_n - D_{n+1}$ which proves the first part. We now assume that the iteration does not stop. We see, that the sequence $\{D_n\}$ is monotonically decreasing and bounded, therefore convergent. We show next, that $\{D_n\}$ is a zero sequence. The very same trick as used below to show convergence may also be found in e.g. [10] and [16]. Then we have

$$D_{n+1}^{1-p} - D_n^{1-p} \geq \left(\beta_n - \frac{\gamma_n}{p^* D_n}\beta_n^{p^*}\right) D_n^{1-p} \geq \frac{1}{p^*}\min\{D_0^{1-p}, \gamma_n^{1-p}\} > 0.$$

If $\gamma_n \leq \Gamma < \infty$ this gives

$$D_n^{1-p} \geq \sum_{k=0}^{n-1} D_{k+1}^{1-p} - D_k^{1-p} \geq \frac{n}{p^*}\min\left\{D_0^{1-p}, \Gamma^{1-p}\right\}$$

and therefore

$$D_n \leq (p^*)^{\frac{1}{p-1}} \max\{D_0, \Gamma\}\, n^{-\frac{1}{p-1}}.$$

This proves the second part with $C := (p^*)^{1/(p-1)}\max\{D_0, \Gamma\}$. Moreover, we have $\Delta_n \to 0$ as $n \to \infty$. By the coercivity of the Bregman distance we conclude $x_n^\delta \to x^\dagger$. Moreover, with the lower bound on γ_n we derive

$$\beta_n \leq \frac{D_n^{p-1}}{\gamma_n^{p-1}} \leq \frac{p^*}{\gamma_n^{p-1}}\max\{D_0, \Gamma\}^{p-1} n^{-1} \leq p^* \max\left\{\left(\frac{D_0}{\gamma}\right)^{p-1}, \left(\frac{\Gamma}{\gamma}\right)^{p-1}\right\} n^{-1}.$$

This proves the last part of the theorem. ■

4.4.3 A convergence and stability result

We now discuss convergence and stability aspects of the algorithm. We start with the noiseles case $\delta = 0$. From the suggested choice of $\bar{R}$ in Algorithm 4.4.1 we see that this situation needs some special treatment. Here we have to distinguish between two cases:

- For fulfilled reference source condition (4.11), we can choose the function $d_0(R)$, $R \geq 0$, such that $d_0(R) \equiv 0$ for $R \geq \|\omega\|$. Then we can set $\infty > \bar{R} \geq \|\omega\|$ and continue as in Algorithm 4.4.1 suggested.
- On the other hand, if $\xi^\dagger \notin \mathcal{R}(F'(x^\dagger)^\star)$ we have $d_0(R) > 0$ for all $R \geq 0$. Hence we cannot find $\bar{R} < \infty$ such that $d_0(\bar{R}) \leq \delta\,\bar{R} = 0$. But we also observe $\bar{R} \to \infty$ as $\delta \to 0$ in that case. It turns out in the subsequent analysis that then $\gamma_n \to \infty$ holds which implies $\beta_n \to 0$ as $\delta \to 0$. Therefore we set $\beta_n \equiv 0$ in this situation. Then the algorithm in principle reduces to the form which was considered in the previous section (with another choice of the step size μ_n). In particular, inequality (4.16) reduces to

$$\Delta_p(x^\dagger, x_{n+1}) \leq \Delta_p(x^\dagger, x_n) - \mu(1-L)\,\|F(x_n) - y\|^p + 2^{p^*-1}\frac{G_{p^*}^*}{p^*}\|\psi_n^*\|^{p^*}\mu^{p^*}$$

for all $\mu > 0$ (of course, then the factor 2^{p^*-1} in the last term can be skipped but is left in the inequality in order to obtain later on a continuous dependency of the choices of μ_n and δ_n on the noise-level δ).

We formulate the following statement. Again we use the notation $x_n := x_n^0$ for the iterates.

Lemma 4.4.3 *Assume (B1)-(B3), $\delta = 0$ and all iterates $\{x_n\}$ remain in $\mathcal{D}(F)$. Let the stopping criterion be chosen such that the iteration only stops after a finite number N of iterations if $F(x_N) = y$. Assume either $\gamma_n \leq \Gamma$ if $\xi^\dagger \in \mathcal{R}(F'(x^\dagger)^\star)$ or $\beta_n \equiv 0$. If the iteration does not stop then we have convergence $x_n \to x^\dagger$ as $\delta \to 0$.*

PROOF. If $\xi^\dagger \in \mathcal{R}(F'(x^\dagger)^\star)$ we can apply Theorem 4.4.1(ii). For $\beta_n \equiv 0$ we can apply the results of Section 4.3. ■

We now suppose $\delta > 0$. Let $N(\delta, y^\delta)$ again denote the index where the iteration stops. We now discuss conditions ensuring $x^\delta_{N(\delta,y^\delta)} \to x^\dagger$ as $\delta \to 0$. We formulate the following general result.

Theorem 4.4.2 *Assume (B1)-(B3) and let the parameters μ_n and β_n depend continuously on y^δ. Moreover, assume that inequality (4.17) holds for all $n \leq N(\delta, y^\delta)$ and let one of the following conditions hold:*

(i) We have $0 < \gamma_n \leq \Gamma < \infty$ uniformly for all $n \leq N(\delta, y^\delta)$ and all $\delta > 0$ or

(ii) for all $\delta > 0$ the following holds:

(ii.a) $\mu_n \leq \overline{\mu} < \infty$ for all $n \leq N(\delta, y^\delta)$,

(ii.b) $C(\mu_n) \geq C_\mu \|F(x_n^\delta) - y^\delta\|^p$ for some constant $C_\mu > 0$ and all $n \leq N(\delta, y^\delta)$,

(ii.c) the sequence $\{\beta_n\}$ remains summable, i.e. the bound $\sum_{n=0}^{N(\delta,y^\delta)} \beta_n \leq \bar{C}_\beta < \infty$ holds uniformly for some constant $\bar{C}_\beta > 0$ and

(ii.d) $N(\delta, y^\delta) \leq C_N \delta^{-p}$ holds for some constant $C_N > 0$.

If all iterates $\{x_n^\delta\}$ remain in $\mathcal{D}(F)$ then we have $x^\delta_{N(\delta,y^\delta)} \to \tilde{x}^\dagger$ with $F(\tilde{x}^\dagger) = y$ as $\delta \to 0$.

PROOF. Let $\{\delta_j\}$ denote a sequence of noise levels with $\delta_j \to 0$ as $j \to \infty$. Without loss of generality we suppose $N(\delta_j, y^{\delta_j}) \to \infty$ as $j \to \infty$. Otherwise we see that the iterates depend continuously on y^{δ_j}. Under assumption (i) the constant in the estimate (4.18) does not depend on δ which proves the first part. In case (ii) we observe with $\delta_j = \delta$

$$\sum_{n=0}^{N(\delta,y^\delta)-1} \|F(x_n^\delta) - y^\delta\|^p \leq \frac{1}{C_\mu} \sum_{n=0}^{N(\delta,y^\delta)-1} C(\mu_n) \leq \frac{D_0}{C_\mu} < \infty$$

which proves $F(x^\delta_{N(\delta,y^\delta)}) \to y$ as $\delta \to 0$. We now continue as in the proof of Theorem 4.3.1. Hence for $l < k \leq N(\delta, y^\delta)$ and $\|F(x_n^\delta) - y^\delta\| \geq \|F(x_k^\delta) - y^\delta\|$ for all $l \leq n \leq k$ we have

$$\begin{aligned}
|\langle J_p(x_k^\delta) - J_p(x_l^\delta), x_k^\delta - x^\dagger\rangle| &= \left|\sum_{n=l}^{k-1}\langle J_p(x_{n+1}^\delta) - J_p(x_n^\delta), x_k^\delta - x^\dagger\rangle\right| \\
&= \left|\sum_{n=l}^{k-1}\langle \mu_n F'(x_n^\delta)^*(F(x_n^\delta) - y^\delta) + \beta_n(x_n^* - x_\sharp^*), x_k^\delta - x^\dagger\rangle\right| \\
&\le \left|\sum_{n=l}^{k-1}\mu_n\langle J_p(F(x_n^\delta) - y^\delta), F'(x_n^\delta)(x_k^\delta - x_n^\delta + x_n^\delta - x^\dagger)\rangle\right| + \left|\sum_{n=l}^{k-1}\beta_n\langle x_n^* - x_\sharp^*, x_k^\delta - x^\dagger\rangle\right| \\
&\le \overline{\mu}\sum_{n=l}^{k-1}\|F(x_n^\delta) - y^\delta\|^{p-1}3(1+L)\left(\delta + \|F(x_n^\delta) - y^\delta\|\right) + \sum_{n=l}^{k-1}\beta_n\|x_n^* - x_\sharp^*\|\,\|x_k^\delta - x^\dagger\| \\
&\le 3\overline{\mu}\,(1+L)\left(\sum_{n=l}^{k-1}\|F(x_n^\delta) - y^\delta\|^p + \sum_{n=l}^{k-1}\|F(x_n^\delta) - y^\delta\|^{p-1}\delta\right) \\
&\qquad + \sum_{n=l}^{k-1}\beta_n\|x_n^* - x_\sharp^*\|\,\|x_k^\delta - x^\dagger\|.
\end{aligned}$$

The first sum tends to zeros as $l \to \infty$. For the second sum we apply Young's inequality in order to derive

$$\begin{aligned}
\sum_{n=l}^{k-1}\|F(x_n^\delta) - y^\delta\|^{p-1}\delta &\le \frac{C_l}{p^*}\sum_{n=l}^{k-1}\|F(x_n^\delta) - y^\delta\|^p + \frac{1}{p}C_l^{-\frac{p}{p^*}}\sum_{n=l}^{k-1}\delta^p \\
&\le \frac{C_l}{p^*}\sum_{n=l}^{k-1}\|F(x_n^\delta) - y^\delta\|^p + \frac{C_N}{p}C_l^{1-p}
\end{aligned}$$

with arbitrary constant $C_l > 0$. Here we additionally used that $k \le N(\delta, y^\delta) \le C_N\delta^{-p}$. We now choose

$$C_l := \left(\sum_{n=l}^{k-1}\|F(x_n^\delta) - y^\delta\|^p\right)^{-\frac{1}{2}}.$$

Then we see that both summands on the right hand side of the last inequality vanish as $l \to \infty$. On the other hand we derive

$$\begin{aligned}
\sum_{n=l}^{k-1}\beta_n\|x_n^* - x_\sharp^*\|\,\|x_k^\delta - x^\dagger\| &\le \varrho\sum_{n=l}^{k-1}\beta_n\left(\|x_n^* - x^\dagger\| + \|x^\dagger - x_\sharp^*\|\right) \\
&\le \varrho(\varrho + \|x^\dagger - x_\sharp^*\|)\sum_{n=l}^{k-1}\beta_n \to 0 \quad \text{as} \quad l \to \infty.
\end{aligned}$$

We used that $x_n^\delta \in \mathcal{B}_\varrho(x^\dagger)$. We can again apply the argumentation of [88] to prove the assertion. ■

Remark 4.4.1 *The condition (ii.c) can be guaranteed easily by a slight modification: in inequality (4.17) we replace* γ_n *by the term* $\max\{\gamma_n, D_n\bar{\beta}_n^{-\frac{1}{p-1}}\}$ *where* $\{\bar{\beta}_n\}$ *is a summable* *sequence. Then, inequality (4.17) still remains valid and condition (ii.c) is satisfied automatically since this modification leads to $\beta_n \le \bar{\beta}_n$. The other conditions of Theorem 4.4.2 have to be ensured by the specific choice of the parameters β_n and μ_n.*

4.4.4 On some stopping criterions

We now discuss three different types of specific stopping critertions. Then the argumentation is always similar. First we show that – as long as the stopping criterion is not fulfilled – we can find constants $C_{n,1}$ $C_{n,2}$ and γ_n such that the inequality (4.17) is valid. This gives the specific choice of the step size μ_n and the parameter β_n in each iteration step. Then we verify the conditions of Theorem 4.4.2 which then ensures us stability and convergence.

As in the last section additionally we use the notation $\psi_n^* := F'(x_n^\delta)^* J_p(F(x_n^\delta) - y^\delta)$ and $\phi_n^* := x_n^* - x_\sharp^*$.

a) Variant I

We present a first stopping criterion directly based on the discrepancy principle:

(STOP1) For given parameter $\tau_\mu > \frac{1+L}{1-L} > 1$ STOP if $\|F(x_n^\delta) - y^\delta\| \le \tau_\mu \delta$.

Moreover we assume that we have no information about the approximate source condition, i.e. we choose $\bar{R} \equiv 0$. Then the following holds:

Corollary 4.4.1 *Assume (B1)-(B3) and all iterates $\{x_n^\delta\}$ remain in $\mathcal{D}(F)$. For given $\overline{\mu} > 0$ we choose the parameter*

$$
\begin{aligned}
C_{n,1} &:= \frac{c_n^\delta}{p^*}, \quad C_{n,2} := \max\left\{2^{p^*-1} G_{p^*}^* \|\psi_n^*\|^{p^*}, \overline{\mu}^{-\frac{1}{p-1}} C_{n,1}\right\} \\
\tilde{\gamma}_n &:= \left(\frac{C_{\xi^\dagger}}{\tau_\mu}\right)^{p^*} \left(\frac{\|F(x_n^\delta) - y^\delta\|}{\mu_n\left((1-L)\|F(x_n^\delta) - y^\delta\| - (1+L)\delta\right)}\right)^{p^*-1} \quad \text{and} \\
\gamma_n &:= \tilde{\gamma}_n \delta^{-p^*} + 2^{p^*-1} G_{p^*}^* \|\phi_n^*\|^{p^*}.
\end{aligned}
$$

Then the following holds true:

(i) For all $\delta > 0$ inequality (4.17) remains valid as long as the stopping criterion (STOP1) is not fulfilled and $0 < \gamma \le \gamma_n \le \Gamma < \infty$ holds for some constants $\gamma, \Gamma > 0$ which do not depend on n.

(ii) For every $\delta > 0$ the algorithm stops after a finite number $N(\delta, y^\delta)$ of iterations. Moreover, we have the estimate

$$
\Delta_p(x^\dagger, x_{N(\delta,y^\delta)}^\delta) \le C\, N(\delta, y^\delta)^{-\frac{1}{p-1}}. \tag{4.18}
$$

for some constant $C > 0$ and $x_{N(\delta,y^\delta)}^\delta \to \tilde{x}^\dagger$ with $F(\tilde{x}^\dagger) = y$ as $\delta \to 0$.

PROOF. Since the iteration did not stop at the last step we have $\tau_\mu \delta \le \|F(x_n^\delta) - y^\delta\|$. We assume $\beta_n, \mu_n > 0$. We set $\bar{R} := 0$. Then we derive

$$\begin{aligned} C_{\xi^\dagger}\beta_n &\leq C_{\xi^\dagger}\frac{\|F(x_n^\delta) - y^\delta\|}{\delta\,\tau_\mu}\left(c_n^\delta\mu_n\right)^{\frac{1}{p}}\beta_n\left(c_n^\delta\mu_n\right)^{-\frac{1}{p}} \\ &\leq \frac{1}{p^*}\left(\frac{C_{\xi^\dagger}}{\tau_\mu}\right)^{p^*}\left(\frac{\|F(x_n^\delta) - y^\delta\|}{\mu_n((1-L)\,\|F(x_n^\delta) - y^\delta\| - (1+L)\,\delta)}\right)^{\frac{1}{p-1}}\delta^{-p^*}\beta_n^{p^*} + \frac{c_n^\delta}{p}\mu_n. \end{aligned}$$

With γ_n and $C_{n,1}$ as given above this gives

$$\Delta_{n+1} \leq (1-\beta_n)\,\Delta_n + \frac{\gamma_n}{p^*}\beta_n^{p^*} - C_{n,1}\mu_n + 2^{p^*-1}\frac{G_{p^*}^*}{p^*}\|\psi_n^*\|^{p^*}\mu_n^{p^*}.$$

We check γ_n. We have

$$\begin{aligned} \mu_n &\geq \left(\frac{c_n^\delta}{2^{p^*-1}p^*G_{p^*}^*\|\psi_n^*\|^{p^*}}\right)^{p-1} \\ &\geq \left(\frac{(1-L)\,\|F(x_n^\delta) - y^\delta\|^p - (1+L)\,\delta\,\|F(x_n^\delta) - y^\delta\|^{p-1}}{2^{p^*-1}p^*G_{p^*}^*K^{p^*}\|F(x_n^\delta) - y^\delta\|^p}\right)^{p-1} \\ &= \left(\frac{1}{2^{p^*-1}p^*G_{p^*}^*K^{p^*}}\left[1 - L - \frac{(1+L)\,\delta}{\|F(x_n^\delta) - y^\delta\|}\right]\right)^{p-1} \\ &> \left(\frac{(1-L)\tau_\mu - (1+L)}{2^{p^*-1}p^*G_{p^*}^*K^{p^*}\tau_\mu}\right)^{p-1} =: \underline{\mu} \end{aligned}$$

since $\delta\,\|F(x_n^\delta) - y^\delta\|^{-1} \leq \tau_\mu^{-1} < (1-L)/(1+L)$. A similar argumentation gives

$$1 > \frac{(1-L)\,\|F(x_n^\delta) - y^\delta\| - (1+L)\delta}{\|F(x_n^\delta) - y^\delta\|} \geq 1 - L - \frac{1+L}{\tau_\mu} = \frac{(1-L)\tau_\mu - (1+L)}{\tau_\mu} > 0.$$

Hence we can define

$$\hat{\Gamma} := \left(\frac{C_{\xi^\dagger}}{\tau_\mu\delta}\right)^{p^*}\left(\frac{\tau_\mu}{\underline{\mu}\,((1-L)\tau_\mu - (1+L))}\right)^{p^*-1} = \frac{C_{\xi^\dagger}^{p^*}}{\tau_\mu}\left(\frac{1}{\underline{\mu}\,((1-L)\tau_\mu - (1+L))}\right)^{p^*-1}\delta^{-p^*}$$

and

$$\Gamma := \hat{\Gamma} + \sup\left\{2^{p^*-1}G_{p^*}^*\|J_p(x) - x_\sharp^*\|^{p^*} \;:\; \Delta_p(x^\dagger, x) \leq D_0\right\},$$

which implies $\gamma_n \leq \Gamma$. The lower bound on γ_n we derive by

$$\gamma_n \geq \left(\frac{C_{\xi^\dagger}}{\tau_\mu}\right)^{p^*}\frac{\delta^{-p^*}}{\mu_n^{p^*-1}} \geq \left(\frac{C_{\xi^\dagger}}{\tau_\mu}\right)^{p^*}\frac{\delta^{-p^*}}{\bar{\mu}^{p^*-1}} =: \gamma.$$

We can now apply Theorem 4.4.1 which proves the first part of the corollary. For the second part check the validity of the conditions of Theorem 4.4.2(ii). We have $\mu_n \leq \bar{\mu}$ by the choice of $C_{n,2}$ automatically. Furthermore, we get

$$\begin{aligned} C(\mu_n) = \frac{\mu_n}{p}C_{n,1} &= \frac{(1-L)\,\|F(x_n^\delta) - y^\delta\|^p - \delta\,(1+L)\,\|F(x_n^\delta) - y^\delta\|^{p-1}}{p^*p}\mu_n \\ &\geq \frac{(1-L)\tau_\mu - (1+L)}{\tau_\mu p^* p}\underline{\mu}\|F(x_n^\delta) - y^\delta\|^p. \end{aligned}$$

Finally for $n < N(\delta, y^\delta)$ we have $\|F(x_n^\delta) - y^\delta\| \geq \tau_\mu \delta$ and hence

$$D_0 > \sum_{n=0}^{N(\delta,y^\delta)-1} \frac{(1-L)\tau_\mu - (1+L)}{\tau_\mu p^* p} \underline{\mu} \|F(x_n^\delta) - y^\delta\|^p \geq \frac{(1-L)\tau_\mu - (1+L)}{p^* p} \underline{\mu} \tau_\mu^{p-1} N(\delta, y^\delta) \delta^p$$

which proves $N(\delta, y^\delta) \leq C_N \delta^{-p}$ for some constant $C_N > 0$. Finally, from the lower bound on γ_n we obtain

$$\beta_n \leq \left(\frac{D_n}{\gamma_n}\right)^{p-1} \leq D_0^{p-1} \left(\frac{\overline{\mu}^{p^*-1}}{C_{\xi^\dagger}^{p^*}} \|F(x_n^\delta) - y^\delta\|^{p^*}\right)^{p-1} = D_0^{p-1} \frac{\overline{\mu}}{C_{\xi^\dagger}^p} \|F(x_n^\delta) - y^\delta\|^p$$

and hence

$$\sum_{n=0}^{N(\delta,y^\delta)-1} \beta_n \leq D_0^{p-1} \frac{\overline{\mu}}{C_{\xi^\dagger}^p} \sum_{n=0}^{N(\delta,y^\delta)-1} \|F(x_n^\delta) - y^\delta\|^p \leq C\, D_0^p$$

uniformly with a constant $C > 0$ which does not depend on δ. ■

The corollary proves that our algorithm together with the discrepancy principle as stopping criterion is a regularization method. However, in this situation we cannot prove convergence rates. On the other hand, for fulfilled source condition (4.11) we can prove the following.

Corollary 4.4.2 *Assume (B1)-(B3) and all iterates $\{x_n^\delta\}$ remain in $\mathcal{D}(F)$. Moreover, let the source condition (4.11) be satisfied with $\bar{R} \geq \|\omega\|$ and the parameter $C_{n,1}$ and $C_{n,2}$ are chosen as suggested in Corollary 4.4.1. We set*

$$\begin{aligned}
\tilde{\gamma}_n &:= \left(\frac{(1+L)(\tau_\mu+1)}{\tau_\mu}\right)^{p^*} \left(\frac{\|F(x_n^\delta) - y^\delta\|}{\mu_n\left((1-L)\,\|F(x_n^\delta) - y^\delta\| - (1+L)\,\delta\right)}\right)^{p^*-1} \quad \text{and} \\
\gamma_n &:= \tilde{\gamma}_n \bar{R}^{p^*} + 2^{p^*-1} G_{p^*}^* \|\phi_n^*\|^{p^*}.
\end{aligned}$$

Then all results of Corollary 4.4.1 remain true.

PROOF. Since we can suppose $d_0(\bar{R}) = 0$ we have

$$\bar{R}(1+L)\left(\|F(x_n^\delta) - y^\delta\| + \delta\right)\beta_n \leq \|F(x_n^\delta) - y^\delta\| \frac{(1+L)(\tau_\mu+1)}{\tau_\mu} \bar{R}\beta_n.$$

The rest is similar to the proof of Corollary 4.4.1. We only have to observe that the bounds γ and Γ do not depend on δ anymore. ■

b) Variant II

In a second variant we want to present a convergence rates result based on the a-priori knowledge of the function $d_0(R)$, $R \geq 0$. Of course, the corresponding results are more of theoretical interest since this function is in general not given. However, the results will show that we can obtain the same convergence rates for this iterative regularization method as in the case of Tikhonov regularization as long as all parameters of the algorithm are chosen in an optimal way.

We assume $\delta > 0$. We present the following stopping criterion:

(STOP2) For given parameter $\tau_\beta > 0$ STOP if $\tau_\beta \beta_n \bar{R} \leq \delta^{p-1}$ and $\beta_n < 1$.

We can prove the following convergence result.

Corollary 4.4.3 *Assume (B1)-(B3) and all iterates $\{x_n^\delta\}$ remain in $\mathcal{D}(F)$. For given $\overline{\mu} > 0$ we choose the parameter*

$$
\begin{aligned}
C_{n,1} &:= \frac{(1-L)\,\|F(x_n^\delta) - y^\delta\|^p}{2}, \qquad C_{n,2} := \max\left\{2^{p^*-1} G_{p^*}^* \|\psi_n^*\|^{p^*}, \overline{\mu}^{-\frac{1}{p-1}} C_{n,1}\right\},\\
\tilde{\gamma}_n &:= (2+L)\, p^* \tau_\beta^{p^*-1} + \left(\frac{2}{\mu_n\,(1-L)}\right)^{p^*-1} (1+L)^{p^*} + \frac{\mu_n}{p-1}\,(1+L)^p \left(\frac{2}{1-L}\right)^{p-1} \tau_\beta^{p^*},\\
\gamma_n &:= \tilde{\gamma}_n \bar{R}^{p^*} + 2^{p^*-1} G_{p^*}^* \|\phi_n^*\|^{p^*}.
\end{aligned}
$$

Then the following holds true:

(i) For all $\delta > 0$ inequality (4.17) remains valid as long as the stopping criterion (STOP2) is not fulfilled and $0 < \gamma \leq \gamma_n \leq \Gamma < \infty$ holds for some constants $\gamma, \Gamma > 0$ which do not depend on n.

(ii) For every $\delta > 0$ the algorithm stops after a finite number $N(\delta, y^\delta)$ of iterations. Moreover, we have the estimate

$$
\Delta_p(x^\dagger, x_{N(\delta,y^\delta)}^\delta) \leq C\,\bar{R}(\delta)\,\delta \tag{4.19}
$$

for some constant $C > 0$. If, in particular, $\bar{R} = \bar{R}(\delta)$ is chosen such that $\bar{R}(\delta)\,\delta \to 0$ as $\delta \to 0$ we have convergence $x_{N(\delta,y^\delta)} \to x^\dagger$ as $\delta \to 0$.

PROOF. We assume, that the iteration did not stop at the last step, i.e. we have $\delta^{p-1} < \tau_\beta \beta_n \bar{R}$. Assume $\beta_n, \mu_n > 0$. We consider inequality (4.16) with $\mu = \mu_n$ and $\beta = \beta_n$. Then we derive

$$
\bar{R}\,(1+L)\,\delta\beta_n + d_0(\bar{R})^{p^*}\beta_n \leq (2+L)\,\tau_\beta^{\frac{1}{p-1}}\,\bar{R}^{p^*}\beta_n^{p^*} = (2+L)\,\tau_\beta^{p^*-1}\bar{R}^{p^*}\beta_n^{p^*}
$$

and

$$
\begin{aligned}
\bar{R}\,(1+L)\,\|F(x_n^\delta) - y^\delta\|\beta_n &\leq \frac{\mu_n\,(1-L)}{2\,p}\|F(x_n^\delta) - y^\delta\|^p\\
&\quad + \frac{1}{p^*}\left(\frac{2}{\mu_n\,(1-L)}\right)^{\frac{p^*}{p}} (1+L)^{p^*}\,\bar{R}^{p^*}\beta_n^{p^*}\\
&= \frac{\mu_n\,(1-L)}{2\,p}\|F(x_n^\delta) - y^\delta\|^p\\
&\quad + \frac{1}{p^*}\left(\frac{2}{\mu_n\,(1-L)}\right)^{p^*-1} (1+L)^{p^*}\,\bar{R}^{p^*}\beta_n^{p^*}.
\end{aligned}
$$

Finally we estimate

$$
\begin{aligned}
\mu_n \delta \,(1+L)\, \|F(x_n^\delta) - y^\delta\|^{p-1} &\le \frac{\mu_n\,(1-L)}{2\,p^*} \|F(x_n^\delta) - y^\delta\|^p + \frac{\mu_n}{p}\left(\frac{2}{1-L}\right)^{\frac{p}{p^*}} (1+L)^p\,\delta^p \\
&\le \frac{\mu_n\,(1-L)}{2\,p^*} \|F(x_n^\delta) - y^\delta\|^p \\
&\quad + \frac{\mu_n}{p}\left(\frac{2}{1-L}\right)^{p-1} (1+L)^p \left(\tau_\beta \bar{R}\right)^{p^*} \beta_n^{p^*}.
\end{aligned}
$$

Then, with γ_n in the assumption, inequality (4.16) leads to

$$
\begin{aligned}
\Delta_{n+1} &\le (1-\beta_n)\,\Delta_n - \mu_n\,(1-L)\,\|F(x_n^\delta) - y^\delta\|^p \left(1 - \frac{1}{2\,p^*} - \frac{1}{2\,p}\right) \\
&\quad + 2^{p^*-1} \frac{G_{p^*}^*}{p^*} \|\psi_n^*\|^{p^*} \mu_n^{p^*} + \frac{\gamma_n}{p^*} \beta_n^{p^*} \\
&= (1-\beta_n)\,\Delta_n + \frac{\gamma_n}{p^*}\beta_n^{p^*} - \left[\frac{\mu_n\,(1-L)}{2} \|F(x_n^\delta) - y^\delta\|^p - 2^{p^*-1} \frac{G_{p^*}^*}{p^*} \|\psi_n^*\|^{p^*} \mu_n^{p^*}\right].
\end{aligned}
$$

Hence we can apply Theorem 4.4.1 with $C_{n,1}$ and $C_{n,2}$ described as above. We only have to show $\gamma_n \le \Gamma < \infty$ for some $\Gamma > 0$. We have by the choice of $C_{n,2}$ that $\mu_n \le \overline{\mu} < \infty$. Moreover,

$$
\begin{aligned}
\mu_n = \left(\frac{(1-L)\,\|F(x_n^\delta) - y^\delta\|^p}{2^{p^*} G_{p^*}^* \|\psi_n^*\|^{p^*}}\right)^{p-1} &\ge \left(\frac{(1-L)\,\|F(x_n^\delta) - y^\delta\|^p}{2^{p^*} G_{p^*}^* \|A_n^*\|\, \|J_p(F(x_n^\delta) - y^\delta)\|^{p^*}}\right)^{p-1} \\
&\ge \left(\frac{1-L}{2^{p^*} G_{p^*}^* K^{p^*}}\right)^{p-1} =: \underline{\mu}
\end{aligned}
$$

Hence we can define

$$
\hat{\Gamma} := (2+L)\,p^*\,\tau_\beta^{p^*-1} \bar{R}^{p^*} + \left(\frac{2}{\underline{\mu}\,(1-L)}\right)^{p^*-1} (1+L)^{p^*} \bar{R}^{p^*} + \frac{\overline{\mu}}{p-1}\left(\frac{2}{1-L}\right)^{p-1} (1+L)^p \left(\tau_\beta \bar{R}\right)^{p^*}
$$

and

$$
\Gamma := \hat{\Gamma} + \sup\left\{2^{p^*-1} G_{p^*}^* \|J_p(x) - x_\sharp^*\|^{p^*} \;:\; \Delta_p(x^\dagger, x) \le D_0\right\},
$$

which implies $\gamma_n \le \Gamma$. Moreover we set $\gamma := 2\,p^*\,\tau_\beta^{p^*-1} \bar{R}^{p^*}$ which gives $0 < \gamma \le \gamma_n$. Hence, by Theorem 4.4.1, $\{\beta_n\}$ is a zero sequence and the stopping index $N(\delta, y^\delta)$ is well-defined. Then we have that $\beta_{N(\delta,y^\delta)} = (D_{N(\delta,y^\delta)}/\gamma_{N(\delta,y^\delta)})^{p-1}$ and $\tau\beta_{N(\delta,y^\delta)}\bar{R} \le \delta^{p-1}$. Further there exists a constant $c = c(\delta_{\max}) > 0$ such that $\bar{R}(\delta) \ge c$ for all $\delta \le \delta_{\max}$. Hence we can find a number $C > 0$ such that $\Gamma \le C\,\bar{R}(\delta)^{p^*}$ for all $\delta \le \delta_{\max}$. Therefore

$$
\Delta_{N(\delta,y^\delta)} \le D_{N(\delta,y^\delta)} \le \gamma_{N(\delta,y^\delta)} \beta_{N(\delta,y^\delta)}^{\frac{1}{p-1}} \le \Gamma\,\tau_\beta^{-\frac{1}{p-1}} \bar{R}^{-\frac{1}{p-1}} \delta \le C\,\tau_\beta^{-\frac{1}{p-1}}\,\delta\,\bar{R}(\delta),
$$

which proves the corollary. ■

We also state the following result which can be considered as an a-priori convergence rate result. We assume that the distance function $d(R; \xi^\dagger - x_\sharp^*)$, $R \ge 0$, is known.

Theorem 4.4.3 *Under the condition of Corollary 4.4.3 we assume that we have chosen* $d_0(R) := C_0 d(R; \xi^\dagger - x_\sharp^*)$, $R \geq 0$, *for some constant* $C_0 > 0$. *Then the following holds:*

(i) If $\xi^\dagger - x_\sharp^* \in \mathcal{R}(F'(x^\dagger)^\star)$, *i.e.* $\xi^\dagger - x_\sharp^* = F'(x^\dagger)^\star \omega$ *for some* $\omega \in \mathcal{Y}^*$, *then the choice of* $\bar{R}$ *with* $\bar{R} \geq \|\omega\|$ *independent of the noise level* δ *leads to the convergence rate*

$$\Delta_p(x^\dagger, x^\delta_{N(\delta,y^\delta)}) \leq C\,\delta. \tag{4.20}$$

(ii) If $\xi^\dagger - x_\sharp^* \in \overline{\mathcal{R}(F'(x^\dagger)^\star)} \setminus \mathcal{R}(F'(x^\dagger)^\star)$ *then the choice* $\bar{R}(\delta) := \Phi^{-1}(\delta)$ *with function* $\Phi(R) := d_0(R)^{p^*} R^{-1}$ *leads to the optimal convergence rate*

$$\Delta_p(x^\dagger, x^\delta_{N(\delta,y^\delta)}) \leq C\,\delta\,\Phi^{-1}(\delta) = C\,d\left(\Phi^{-1}(\delta)\right)^{p^*}. \tag{4.21}$$

It turns out that we end at the same convergence rates results as for Tikhonov regularization in the previous chapter.

Remark 4.4.2 *We compare these rates with the convergence rates results for Tikhonov regularization in Section 3.2. Using the penalty term* $P(x) := \frac{1}{p}\|x\|^p$, *the relation* $p = s$, *as well as the degree of nonlinearity* $(1,0)$, *i.e.* $\kappa = c_1 = 1$ *and* $c_2 = 0$ *then the rates (4.20) and (4.21) coincides with the convergence rates of Theorem 3.2.1 and Theorem 3.2.3 respectively.*

The results of Corollary 4.4.3 do not provide convergence when we have chosen $\bar{R} := C_{\xi^\dagger}\delta^{-1}$. Here, (4.19) represents only a boundedness result. We recall that the underlying choice of the parameter $\bar{R}$ can be considered as the case where no a-priori information on $\xi^\dagger - x_\sharp^*$ is applied. Here we state the following lemma.

Lemma 4.4.4 *Suppose (B1)-(B3). Assume that* $\bar{R} := C_{\xi^\dagger}\delta^{-1}$ *and let the parameters* $C_{n,1}$ *and* $C_{n,2}$ *be chosen according to Corollary 4.4.3 whereas* γ_n *is replaced by* $\max\{\gamma_n, D_n \bar{\beta}_n^{-\frac{1}{p-1}}\}$ *with nonnegative sequence* $\{\bar{\beta}\}$ *satisfying* $\sum\limits_{n=0}^{\infty} \bar{\beta}_n < \infty$. *Then we have* $x^\delta_{N(\delta,y^\delta)} \to x^\dagger$ *as* $\delta \to 0$.

Proof. We check the validity of the conditions of Theorem 4.4.2(ii). We have $\mu_n \leq \bar{\mu}$ by the choice of $C_{n,1}$ automatically. Furthermore, from the proof of Corollary 4.4.3 we get

$$C(\mu_n) = \frac{1}{p}\mu_n C_{n,1} \geq \frac{\mu(1-L)}{2p}\|F(x_n^\delta) - y^\delta\|^p.$$

The third part is given by the additional assumption. Let $\{\tilde{\beta}_n\}$ denote the parameters given by Corollary 4.4.3. Then the modification of γ_n implies $\beta_n := \min\{\bar{\beta}_n, \tilde{\beta}_n\}$. From the proof of Corollary 4.4.3 we see that $\tilde{\beta}_n \leq C_\beta n^{-1}$ with constant C_β not depending on δ. From the stopping criterion we now derive for all $n < N(\delta, y^\delta)$ that $\beta_n \leq C_\beta n^{-1}$ and hence

$$\delta^{p-1} \leq \tau_\beta \beta_n \bar{R} \leq \tau_\beta \tilde{\beta}_n \bar{R} = \tau_\beta C_\beta n^{-1} C_{\xi^\dagger}\delta^{-1}$$

or equivalently $n \leq \tau_\beta C_\beta C_{\xi^\dagger} \delta^{-p}$. This proves the last assumption by choosing $n = N(\delta, y^\delta) - 1$. ■

In the algorithm the role of the parameter $\tau_\beta > 0$ was left open. In order to understand the influence of the specific choice of τ_β in the algorithm we consider the limit situations $\tau_\beta \longrightarrow 0$ and $\tau_\beta \longrightarrow \infty$ keeping all other parameters constant:

- Assume $\tau_\beta \longrightarrow 0$. Then we see that $\gamma_n \sim$ *const.* and $\Gamma \sim$ *const.* which provides $\beta_n \sim \gamma_n^{1-p} \sim$ *const.* For the stopping condition we conclude $\tau_\beta \beta_n \bar{R} \longrightarrow 0$ as $\tau_\beta \longrightarrow 0$. Moreover, for the error estimate (4.19) we have $\Delta_p(x^\dagger, x^\delta_{N(\delta,y^\delta)}) \leq C(\tau_\beta)\, \delta\, \bar{R}$ with

$$C(\tau_\beta) = C\,\Gamma\, \tau_\beta^{-\frac{1}{p-1}} \sim \tau_\beta^{-\frac{1}{p-1}} \longrightarrow \infty \quad \text{as} \quad \tau_\beta \longrightarrow 0.$$

- Assume now $\tau_\beta \longrightarrow \infty$. Then we observe $\gamma_n \sim \tau_\beta^{p^*} = \tau_\beta^{\frac{p}{p-1}}$ and $\Gamma \sim \tau_\beta^{\frac{p}{p-1}}$ and hence $\beta_n \sim \gamma_n^{1-p} \sim \tau_\beta^{-p} \longrightarrow 0$ as $\tau_\beta \longrightarrow \infty$. This leads again to $\tau_\beta \beta_n \bar{R} \sim \tau_\beta^{1-p} \longrightarrow 0$ as $\tau_\beta \longrightarrow \infty$. For the error bound we end at

$$C(\tau_\beta) \sim \tau_\beta^{\frac{p}{p-1}}\, \tau_\beta^{-\frac{1}{p-1}} = \tau_\beta \longrightarrow \infty \quad \text{as} \quad \tau_\beta \longrightarrow \infty.$$

So this shows that τ_β should be chosen neither too small nor too large. On the other hand the following observation was made in numerical tests: choosing the parameter τ_β moderate, i.e. $\tau_\beta \in [1, 10]$, the algorithm stops too late with a non-satisfying solution $x^\delta_{N(\delta,y^\delta)}$. A sufficient increase or decrease of the parameter τ_β in fact leads to a smaller number of iteration. However, a parameter τ_β chosen too small or too large might let the iteration stop too early with a worse result. Since the 'optimal' parameter τ_β is not known a priori we therefore suggest an adaptive choice of the parameter τ_β which stands the test in practical applications:

1. At the beginning choose τ_β in a moderate way such that the iteration does not stop too early.
2. If some additional information indicates that the current iterate x_n^δ represents a satisfactory solution we slow down the parameter τ_β successively. This reduces the stopping bound $\tau_\beta \beta_n \bar{R}$ which forces the algorithm to an earlier termination. Such additional information might be for example the discrepancy principle.

We also discuss an alternative approach for the choice of the step size parameters which includes a discrepancy-like criterion. Assume $\|F(x_n^\delta) - y^\delta\| \leq \tau_\mu \delta$ for some $n > 0$ and some $\tau_\mu > \frac{1+L}{1-L}$. Then, as long as the stopping criterion (STOP2) is not fulfilled we have

$$\mu_n \delta\, (1+L)\, \|F(x_n^\delta) - y^\delta\|^{p-1} \leq \mu_n (1+L)\, \tau_\mu^{p-1} \delta^p \leq (1+L) \mu_n \tau_\mu^{p-1} \tau_\beta^{p^*} \bar{R}^{p^*} \beta_n^{p^*}$$

and

$$\bar{R}\, (1+L)\, \|F(x_n^\delta) - y^\delta\|\, \beta_n \leq \tau_\mu (1+L)\, \delta\, \bar{R}\, \beta_n \leq (1+L)\, \tau_\mu \tau_\beta^{p^*-1} \bar{R}^{p^*} \beta_n^{p^*}.$$

Therefore we suggest the following choice of the parameters:

$$\begin{aligned}
C_{n,1} &:= \begin{cases} \dfrac{(1-L)\,\|F(x_n^\delta)-y^\delta\|^p}{2}, & \|F(x_n^\delta)-y^\delta\| > \tau_\mu\delta, \\ (1-L)\,\|F(x_n^\delta)-y^\delta\|^p, & \|F(x_n^\delta)-y^\delta\| \le \tau_\mu\delta, \end{cases} \\
C_{n,2} &:= \max\left\{2^{p^*-1}G^*_{p^*}\|\psi_n^*\|^{p^*}, \overline{\mu}^{-\frac{1}{p-1}}C_{n,1}\right\}, \\
\tilde{\gamma}_n &:= \begin{cases} (2+L)\,p^*\,\tau_\beta^{p^*-1} + \left(\dfrac{2}{\mu_n(1-L)}\right)^{p^*-1}(1+L)^{p^*} \\ \qquad + \dfrac{\mu_n}{p-1}\left(\dfrac{2}{1-L}\right)^{p-1}(1+L)^p\tau_\beta^{p^*}, & \|F(x_n^\delta)-y^\delta\| > \tau_\mu\delta, \\ p^*\left((2+L)\,\tau_\beta^{p^*-1} + \mu_n(1+L)\,\tau_\mu^{p-1}\tau_\beta^{p^*} \right. \\ \qquad \left. + (1+L)\,\tau_\mu\tau_\beta^{p^*-1}\right), & \|F(x_n^\delta)-y^\delta\| \le \tau_\mu\delta, \end{cases} \\
\gamma_n &:= \tilde{\gamma}_n\bar{R}^{p^*} + 2^{p^*-1}G^*_{p^*}\|\phi_n^*\|^{p^*}.
\end{aligned}$$

The consequence is simple: as long as the step size μ_n is not too small the modified choice leads to a larger γ_n when the discrepancy criterion is fulfilled. This provides a smaller parameter β_n which leads to a earlier termination of the iteration. Hence the discrepancy principle influences the stopping index $N(\delta, y^\delta)$ of the iteration. It is a short check to observe, that the results of Corollary 4.4.3 and Theorem 4.4.3 remain valid.

c) Variant III

In a last idea we combine both stopping criterions:

(STOP3) For given parameter $\tau_\mu > \frac{1+L}{1-L} \ge 1$ and $\tau_\beta > 0$ STOP if either $\|F(x_n^\delta) - y^\delta\| \le \tau_\mu\delta$ or $\tau_\beta\beta_n^{\frac{1}{p-1}} \le \delta^{p^*+1}\bar{R}$ and $\beta_n < 1$.

We present the following result.

Corollary 4.4.4 *Assume (B1)-(B4) and all iterates $\{x_n^\delta\}$ remain in $\mathcal{D}(F)$. For given $\overline{\mu} > 0$ we choose the parameter*

$$\begin{aligned}
C_{n,1} &:= \frac{c_n^\delta}{p^*}, \quad C_{n,2} := \max\left\{2^{p^*-1}G^*_{p^*}\|\psi_n^*\|^{p^*}, \overline{\mu}^{-\frac{1}{p-1}}C_{n,1}\right\}, \\
\tilde{\gamma}_n &:= C_{\xi^\dagger}^{p^*}\left(\frac{\|F(x_n^\delta)-y^\delta\|}{\mu_n\left((1-L)\,\|F(x_n^\delta)-y^\delta\| - (1+L)\,\delta\right)}\right)^{p^*-1}
\end{aligned}$$

and either

$$\gamma_n := \left(\frac{2+L+(1+L)\,\tau_\mu}{\tau_\mu}\right)^{p^*}\tilde{\gamma}_n\delta^{-p^*} + 2^{p^*-1}G^*_{p^*}\|\phi_n^*\|^{p^*}$$

or

$$\gamma_n := \left((2+L)\,p^*\,\tau_\beta + \tilde{\gamma}_n\right)\delta^{-p^*} + 2^{p^*-1}G^*_{p^*}\|\phi_n^*\|^{p^*}.$$

Then the following holds true:

(i) For all $\delta > 0$ inequality (4.17) remains valid as long as the stopping criterion (STOP3) is not fulfilled and $0 < \gamma \le \gamma_n \le \Gamma < \infty$ holds for some constants $\gamma, \Gamma > 0$ which do not depend on n.

(ii) For every $\delta > 0$ the algorithm stops after a finite number $N(\delta, y^\delta)$ of iterations. Moreover, we have the estimate

$$\Delta_p(x^\dagger, x^\delta_{N(\delta,y^\delta)}) \le C\, N(\delta, y^\delta)^{-\frac{1}{p-1}}.$$

for some constant $C > 0$ and $x^\delta_{N(\delta,y^\delta)} \to \tilde{x}^\dagger$ with $F(\tilde{x}^\dagger) = y$ as $\delta \to 0$.

(iii) If additionally we have $\tau_\beta \beta^{\frac{1}{p-1}}_{N(\delta,y^\delta)} \le \delta^{p^*+1} \bar{R}$ then the estimate

$$\Delta_p(x^\dagger, x^\delta_{N(\delta,y^\delta)}) \le \tilde{C}\, \bar{R}(\delta)\, \delta.$$

holds for some constant $\tilde{C} > 0$ which does not depend on δ.

Proof. Since the iteration did not stop in the last iteration both conditions are violated, i.e we have $\delta < \tau_\mu^{-1} \|F(x_n^\delta) - y^\delta\|$ and $\delta < \tau_\beta \beta_n^{\frac{1}{p-1}} \bar{R}^{-1} \delta^{-p^*}$. Then we derive

$$\begin{aligned} I &:= \bar{R}\,(1+L)\, \|F(x_n^\delta) - y^\delta\|\, \beta_n + \bar{R}\,\delta\,(1+L)\,\beta_n + d_0(\bar{R})^{p^*} \beta_n \\ &\le \bar{R}\,(1+L)\, \|F(x_n^\delta) - y^\delta\|\, \beta_n + (2+L)\, \bar{R}\, \beta_n \min\left\{ \tau_\mu^{-1} \|F(x_n^\delta) - y^\delta\|, \tau_\beta \beta_n^{\frac{1}{p-1}} \bar{R}^{-1} \delta^{-p^*} \right\}. \end{aligned}$$

Taking the first estimate we follow the argumentation of Corollary 4.4.1. We have $\bar{R} \le C_{\xi^\dagger} \delta^{-1}$ and hence

$$\begin{aligned} I &\le C_{\xi^\dagger} \delta^{-1} \beta_n \left(\frac{2+L}{\tau_\mu} + 1 + L \right) \|F(x_n^\delta) - y^\delta\| \\ &\le \frac{1}{p^*} \left(\frac{(1+L)\tau_\mu + 2 + L}{\tau_\mu} \right)^{p^*} \left(\frac{\|F(x_n^\delta) - y^\delta\|}{\mu_n((1-L)\, \|F(x_n^\delta) - y^\delta\| - (1+L)\,\delta)} \right)^{\frac{1}{p-1}} C_{\xi^\dagger}^{p^*} \delta^{-p^*} \beta_n^{p^*} \\ &\quad + \frac{1}{p} \mu_n c_n^\delta. \end{aligned}$$

On the other hand, for the second estimate we see that

$$\begin{aligned} I &\le C_{\xi^\dagger} \delta^{-1} \beta_n (1+L)\, \|F(x_n^\delta) - y^\delta\| + (2+L)\, \tau_\beta \delta^{-p^*} \beta_n^{p^*} \\ &\le \left[(2+L)\, \tau_\beta + \frac{(1+L)^{p^*} C_{\xi^\dagger}^{p^*}}{p^*} \left(\frac{\|F(x_n^\delta) - y^\delta\|}{\mu_n\,((1-L)\, \|F(x_n^\delta) - y^\delta\| - (1+L)\,\delta)} \right)^{\frac{1}{p-1}} \right] \delta^{-p^*} \beta_n^{p^*} \\ &\quad + \frac{1}{p} \mu_n c_n^\delta. \end{aligned}$$

Hence we set γ_n as suggested in this case which gives

$$\Delta_{n+1} \le (1-\beta_n)\, \Delta_n + \frac{\gamma_n}{p^*} \beta_n^{p^*} + 2^{p^*-1} \frac{G_{p^*}^*}{p^*} \|\psi_n^*\|^{p^*} \mu_n^{p^*} - \frac{c_n^\delta}{p^*} \mu_n.$$

The rest of the proof is the same as in the proofs of Corollary 4.4.1 and 4.4.3. We only have to observe, that

$$\gamma_n \ge \min\left\{ (2+L)\, p^*\, \tau_\beta, \left(\frac{2+L+(1+L)\,\tau_\mu}{\tau_\mu} \right)^{p^*} \frac{(1+L)^{p^*} C_{\xi^\dagger}^{p^*}}{\overline{\mu}^{\frac{1}{p-1}}} \right\} \delta^{-p^*} =: \gamma > 0$$

and $\gamma_n \leq C\,\delta^{-p^*}$ for some constant $C > 0$ for all $n > 0$ as long as the stopping criterion is not fulfilled. Hence, if $\tau_\beta \beta_{N(\delta,y^\delta)}^{\frac{1}{p-1}} \leq \delta^{p^*+1}\bar{R}$ we derived

$$\Delta_p(x^\dagger, x^\delta_{N(\delta,y^\delta)}) \leq D_{N(\delta,y^\delta)} = \gamma_{N(\delta,y^\delta)}\beta_{N(\delta,y^\delta)}^{\frac{1}{p-1}} \leq C\,\tau_\beta^{-1}\delta\,\bar{R},$$

which proves the accordant convergence rates result. ■

Finally we again suggest some alternative choices of the parameters $C_{n,1}$, $C_{n,2}$ and γ_n without violating inequality (4.17).

Remark 4.4.3 *We present some ideas for modified choices of the parameters.*

- *We can improve the estimates in the proof by distinguishing between* $\tau_\mu^{-1}\|F(x_n^\delta) - y^\delta\| \leq \tau_\beta \beta_n^{\frac{1}{p-1}}\bar{R}^{-1}\delta^{-p^*}$ *and* $\tau_\mu^{-1}\|F(x_n^\delta) - y^\delta\| \geq \tau_\beta \beta_n^{\frac{1}{p-1}}\bar{R}^{-1}\delta^{-p^*}$ *using the first variant for* γ_n *of Corollary 4.4.4 in the first and the second in the latter case.*
- *In the case* $\tau_\mu^{-1}\|F(x_n^\delta) - y^\delta\| \leq \tau_\beta \beta_n^{\frac{1}{p-1}}\bar{R}^{-1}\delta^{-p^*}$ *we can also estimate*

$$\begin{aligned} I &\leq \bar{R}\,\beta_n \left(\frac{2 + L + (1+L)\,\tau_\mu}{\tau_\mu}\right)\|F(x_n^\delta) - y^\delta\| \\ &\leq (2 + L + (1+L)\,\tau_\mu)\,\tau_\beta\delta^{-p^*}\beta_n^{p^*}. \end{aligned}$$

This simplifies the parameter choice to

$$\begin{aligned} C_{n,1} &:= c_n^\delta \quad \text{and} \\ \gamma_n &:= p^*\,(2 + L + (1+L)\,\tau_\mu)\,\tau_\beta\delta^{-p^*} + 2^{p^*-1}G^*_{p^*}\|\phi_n^*\|^{p^*}, \end{aligned}$$

whereas $C_{n,2}$ *is chosen as suggested in Corollary 4.4.4. For* $\tau_\mu^{-1}\|F(x_n^\delta) - y^\delta\| > \tau_\beta \beta_n^{\frac{1}{p-1}}\bar{R}^{-1}\delta^{-p^*}$ *we do not change the parameter choices.*

- *A further simplified version of the choice of the parameters is given by the estimates*

$$-\mu_n c_n^\delta \leq -\mu_n \left(\frac{(1-L)\,\tau_\mu - (1+L)}{\tau_\mu}\right)\|F(x_n^\delta) - y^\delta\|^p$$

and

$$\begin{aligned} &C_{\xi^\dagger}\delta^{-1}\beta_n\left(\frac{2 + L + (1+L)\,\tau_\mu}{\tau_\mu}\right)\|F(x_n^\delta) - y^\delta\| \\ &\leq \frac{\mu_n}{p}\left(\frac{(1-L)\,\tau_\mu - (1+L)}{\tau_\mu}\right)\|F(x_n^\delta) - y^\delta\|^p \\ &\quad + \frac{1}{p^*}\left(\frac{\tau_\mu}{\mu_n\left((1-L)\,\tau_\mu - (1+L)\right)}\right)^{\frac{1}{p-1}}\left(\frac{2 + L + (1+L)\,\tau_\mu}{\tau_\mu}\right)^{p^*} C_{\xi^\dagger}^{p^*}\delta^{-p^*}\beta_n^{p^*} \end{aligned}$$

which gives the alternative choices

$$\begin{aligned} C_{n,1} &:= \frac{1}{p^*}\left(\frac{(1-L)\,\tau_\mu - (1+L)}{\tau_\mu}\right)\|F(x_n^\delta) - y^\delta\|^p \qquad \text{and} \\ \gamma_n &:= \frac{1}{\tau_\mu}\frac{(2 + L + (1+L)\,\tau_\mu)^{p^*}}{\left(\mu_n\left((1-L)\,\tau_\mu - (1+L)\right)\right)^{p^*-1}}C_{\xi^\dagger}^{p^*}\delta^{-p^*} + 2^{p^*-1}G^*_{p^*}\|\phi_n^*\|^{p^*}, \end{aligned}$$

which do not distinguish between $\tau_\mu^{-1}\|F(x_n^\delta)-y^\delta\| \le \tau_\beta \beta_n^{\frac{1}{p-1}} \bar{R}^{-1}\delta^{-p^*}$ *and* $\tau_\mu^{-1}\|F(x_n^\delta)-y^\delta\| \ge \tau_\beta \beta_n^{\frac{1}{p-1}} \bar{R}^{-1}\delta^{-p^*}$ *anymore.*

It is not clear a priorly which implementation of the parameter choice is the most efficient one. It might also depend on the specific operator A. It it has to be tesred in the accordant situation which implementation is the best one.

4.4.5 On a further accelerated version

The choices for the parameters β_n and μ_n in the last section can be done explicitely. This advantage is paid by the price of some extra estimations leading to an only suboptimal parameter choices. Now we want to present an improved choice of the step size μ_n. Moreover, as opposite to the previous approaches we now first calculate β_n (explicitely) and define then the step size μ_n as solution of an appropriate minimization problem.

Let therefore $x_n^\delta \in \mathcal{D}(F)$ be given. For arbitrary $\mu, \beta \ge 0$ we derive

$$\begin{aligned}
\Delta_{\mu,\beta} - \Delta_n &= \frac{1}{p^*}\|x_n^* - \beta\phi_n^* - \mu\psi_n^*\|^{p^*} - \frac{1}{p^*}\|x_n^*\|^{p^*} + \mu\langle\psi_n^*, x^\dagger\rangle + \beta\langle\phi_n^*, x^\dagger\rangle \\
&= \frac{1}{p^*}\|x_n^* - \beta\phi_n^* - \mu\psi_n^*\|^{p^*} - \frac{1}{p^*}\|x_n^*\|^{p^*} + \mu\langle\psi_n^*, x_n^*\rangle \\
&\quad -\mu\langle\psi_n^*, x_n^* - x^\dagger\rangle + \beta\langle\phi_n^*, x^\dagger\rangle \\
&\le \frac{1}{p^*}\|x_n^* - \beta\phi_n^* - \mu\psi_n^*\|^{p^*} - \frac{1}{p^*}\|x_n^*\|^{p^*} + \mu\langle\psi_n^*, x_n^*\rangle + \beta\langle\phi_n^*, x^\dagger\rangle - \mu c_n^\delta
\end{aligned}$$

with c_n^δ given by (4.8) We now suppose $\delta > 0$ and assume that the iteration is terminated along the stopping criterion (STOP1). Since the iteration does not stop we have

$$\|F(x_n^\delta) - y^\delta\| > \tau_\mu\delta \quad \text{with} \quad \tau_\mu > \frac{1+L}{1-L}. \tag{4.22}$$

Then we derive

$$c_n^\delta > \frac{(1-L)\tau_\mu - (1+L)}{\tau_\mu}\|F(x_n^\delta) - y^\delta\|^p > ((1-L)\tau_\mu - (1+L))\,|F(x_n^\delta) - y^\delta\|^{p-1}\delta > 0.$$

Assume now that we have β_n already chosen. Then we can define

$$f_{\beta_n}(\mu) := \frac{1}{p^*}\|x_n^* - \beta_n\phi_n^* - \mu\psi_n^*\|^{p^*} + \mu\langle\psi_n^*, x_n^\delta\rangle - \mu\, c_n^\delta.$$

We additionally suppose

$$\beta_n^{p^*-1} \le C_\beta \frac{\delta}{\|\phi_n^*\|^{p^*-1}} \quad \text{with} \quad C_\beta \le \frac{((1-L)\tau_\mu - (1+L))\,\delta}{G_{p^*}^* K}. \tag{4.23}$$

Then we easily can prove the following lemma.

Lemma 4.4.5 *Assume (B1)-(B3) and* $\psi_n^* \neq 0$. *Then the minimization problem*

$$f_{\beta_n}(\mu) \to \min \qquad \text{subject to} \quad \mu > 0 \tag{4.24}$$

has a unique solution $\mu^* > 0$ *as long as (4.22) and (4.23) hold.*

PROOF. Differentiating $f_{\beta_n}(\mu)$ we see

$$f'_{\beta_n}(\mu) = -\langle J^*_{p^*}(x_n^* - \beta_n \phi_n^* - \mu\, \psi_n^*), \psi_n^* \rangle + \langle x_n^\delta, \psi_n^* \rangle - c_n^\delta.$$

By monotonicity of the duality mappings this function $f'_{\beta_n}(\mu)$ is strictly increasing. We have to show $f'_{\beta_n}(0) < 0$. Here we have

$$\begin{aligned} f'_{\beta_n}(0) &= -\langle J^*_{p^*}(x_n^* - \beta_n \phi_n^*), \psi_n^* \rangle + \langle x_n^\delta, \psi_n^* \rangle - c_n^\delta \\ &= -\langle J^*_{p^*}(x_n^*) - J^*_{p^*}(x_n^* - \beta_n \phi_n^*), \psi_n^* \rangle - c_n^\delta. \end{aligned}$$

Moreover we can estimate

$$\begin{aligned} \left| \langle J^*_{p^*}(x_n^*) - J^*_{p^*}(x_n^* - \beta_n \phi_n^*), \psi_n^* \rangle \right| &\leq \| J^*_{p^*}(x_n^*) - J^*_{p^*}(x_n^* - \beta_n \phi_n^*) \| \, \|\psi_n^*\| \\ &\leq G^*_{p^*} \beta_n^{p^*-1} \|\phi_n^*\|^{p^*-1} K \, \|F(x_n^\delta) - y^\delta\|^{p-1}. \end{aligned}$$

Hence, $f'_{\beta_n}(0) < 0$ if

$$\beta_n^{p^*-1} < \frac{c_n^\delta}{G^*_{p^*} \|\phi_n^*\|^{p^*-1} K \, \|F(x_n^\delta) - y^\delta\|^{p-1}} = \frac{(1-L)\|F(x_n^\delta) - y^\delta\| - (1+L)\delta}{G^*_{p^*} \|\phi_n^*\|^{p^*-1} K},$$

which is fulfilled under conditions (4.22) and (4.23). By continuity of $f'_{\beta_n}(\mu)$ there exists a unique element $\mu = \mu^* > 0$ satisfying the necessary optimality condition $f'_{\beta_n}(\mu) = 0$. ■

Consequently, we can present the following variant of Algorithm 4.4.1.

Algorithm 4.4.2 *In Algorithm 4.4.1 we replace step (S1) by*

(S1(a)) Calculate $\psi_n^* := F'(x_n^\delta)^* J_p(F(x_n^\delta) - y^\delta)$ *and* $\phi_n^* := x_n^* - x_\natural^*$. *Compute*

$$\begin{aligned} C_{n,1} &:= \frac{c_n^\delta}{p^*}, \quad C_{n,2} := \max\left\{ 2^{p^*-1} G^*_{p^*} \|\psi_n^*\|^{p^*}, C_{n,1} \overline{\mu}^{1-p} \right\}, \quad \tilde{\mu}_n := \left(\frac{C_{n,1}}{C_{n,2}} \right)^{p-1}, \\ \tilde{\gamma}_n &:= \left(\frac{C_{\xi^\dagger}}{\tau_\mu} \right)^{p^*} \left(\frac{\|F(x_n^\delta) - y^\delta\|}{\mu_n \left((1-L) \, \|F(x_n^\delta) - y^\delta\| - (1+L)\, \delta \right)} \right)^{p^*-1}, \\ \gamma_n &:= \max\left\{ \tilde{\gamma}_n, \left(\frac{D_n}{C_\beta} \right)^{p^*} \left(p^* \left(p G^*_{p^*} \right)^{p^*-1} 2^{\frac{p^*-1}{p-1}} \right)^{-1} \right\} \delta^{-p^*} + 2^{p^*-1} G^*_{p^*} \|\phi_n^*\|^{p^*}, \\ \beta_n &:= \min\left\{ 1, \left(\frac{D_n}{\gamma_n} \right)^{p-1} \right\} \quad \text{and} \\ D_{n+1} &:= (1-\beta_n)\, D_n - \frac{\tilde{\mu}_n}{p} C_{n,1} + \frac{\gamma_n}{p^*} \beta_n^{p^*}. \end{aligned}$$

STOP, if the stopping criterion (STOP1) is fulfilled.

(S1(b)) Find the solution μ^* *of the equation*

$$f'_{\beta_n}(\mu) = 0, \qquad \mu \geq 0. \tag{4.25}$$

Set $\mu_n := \min\{\mu^*, \overline{\mu}\}$.

Then we easily can prove the following result.

Theorem 4.4.4 *Assume (B1)-(B3) and all iterates* $\{x_n^\delta\}$ *remain in* $\mathcal{D}(F)$. *Then, for every* $\delta > 0$ *the algorithm stops after a finite number* $N(\delta, y^\delta)$ *of iterations. Moreover we have convergence* $x_{N(\delta,y^\delta)} \to \tilde{x}^\dagger$ *with* $F(\tilde{x}^\dagger) = y$ *as* $\delta \to 0$.

PROOF. Assume that the iteration did not stop. Then condition (4.22) is automatically satisfied. We set

$$\tilde{C}_n := \left(\frac{D_n}{C_\beta}\right)^{p^*} \left(p^* \left(pG_{p^*}^*\right)^{p^*-1} 2^{\frac{p^*-1}{p-1}}\right)^{-1}.$$

An estimate of γ_n gives

$$\begin{aligned}
\gamma_n &\geq \tilde{C}\delta^{-p^*} + 2^{p^*-1} G_{p^*}^* \|\phi_n^*\|^{p^*} \\
&= \frac{1}{p^*}\left(p^{*\frac{1}{p^*}} \tilde{C}^{\frac{1}{p^*}} \delta^{-1}\right)^{p^*} + \frac{1}{p^*}\left((pG_{p^*}^*)^{\frac{1}{p}} 2^{\frac{p^*-1}{p}} \|\phi_n^*\|^{p^*-1}\right)^p \\
&\geq \left(p^{*\frac{1}{p^*}} \tilde{C}^{\frac{1}{p^*}} (pG_{p^*}^*)^{\frac{1}{p}} 2^{\frac{p^*-1}{p}}\right)\delta^{-1}\|\phi_n^*\|^{p^*-1} = \frac{D_n}{C_\beta}\delta^{-1}\|\phi_n^*\|^{p^*-1}
\end{aligned}$$

which gives

$$\beta_n^{p^*-1} = \frac{D_n}{\gamma_n} \leq C_\beta \frac{\delta}{\|\phi_n^*\|^{p^*-1}}$$

which proves the validity of (4.23).

The choice of $C_{n,1}$ and $C_{n,2}$ is the same as in Corollary 4.4.1. From the proof there we conclude

$$\overline{\mu} \geq \tilde{\mu} > \left(\frac{(1-L)\tau_\mu - (1+L)}{2^{p^*-1} p^* G_{p^*}^* K^{p^*} \tau_\mu}\right)^{p-1} = \underline{\mu}$$

as well as

$$\gamma_n \geq \frac{C_{\xi^\dagger}^{p^*}}{\tau_\mu^{p^*} \overline{\mu}^{p^*-1}} \delta^{-p^*} \quad \text{and} \quad \frac{\tilde{\mu}_n}{p} C_{n,1} \geq \frac{(1-L)\tau_\mu - (1+L)}{\tau_\mu p^* p} \underline{\mu} \|F(x_n^\delta) - y^\delta\|^p.$$

Additionally we observe that

$$\hat{\Gamma} := \max\left\{\frac{C_{\xi^\dagger}^{p^*}}{\tau_\mu}\left(\frac{1}{\underline{\mu}\left((1-L)\tau_\mu - (1+L)\right)}\right)^{\frac{1}{p-1}}, \tilde{C}\right\}\delta^{-p^*}$$

and

$$\Gamma := \hat{\Gamma} + \sup\left\{2^{p^*-1} G_{p^*}^* \|J_p(x) - x_\sharp^*\|^{p^*} \,:\, \Delta_p(x^\dagger, x) \leq D_0\right\},$$

imply $\gamma_n \leq \Gamma$. By the choice of β_n and μ_n we obtain

$$\begin{aligned}\Delta_{n+1} &= \Delta_{\mu_n,\beta_n}\\ &\leq \Delta_n + \frac{1}{p^*}\|x_n^*\|^{p^*} + \beta_n\langle\phi_n^*, x^\dagger\rangle + f_{\beta_n}(\mu_n)\\ &\leq \Delta_n + \frac{1}{p^*}\|x_n^*\|^{p^*} + \beta_n\langle\phi_n^*, x^\dagger\rangle + f_{\beta_n}(\tilde{\mu}_n)\\ &\leq \max\{1-\beta_n, 0\}\, D_n - C(\tilde{\mu}_n) + \frac{\gamma_n}{p^*}\beta_n^{p^*} = D_{n+1}\end{aligned}$$

which proves $\Delta_{n+1} \leq D_{n+1}$. Moreover, this shows that we can follow the lines of the proof of Corollary 4.4.1 for verifying the conditions of Theorem 4.4.2(ii). This completes the proof. ■

4.4.6 A numerical study

In this numerical experiments we return to the same situation which was considered in Section 4.3.4. The numerical results are presented in Table 4.3 for $x_1^\dagger$ and in Table 4.4 for the second function $x_2^\dagger$. Additionally we set

$$\tau_\mu := 1.2, \; C_{\xi^\dagger} := 10 \quad \text{and} \quad D_0 := 100.$$

Summarizing the numerical results we can state the following:

- Even not presented here the additional regularization term does not yield to additional regularization errors. There is no significant difference in the error of the regularized solutions obtained by Landweber respectively modified Landweber iteration methods.
- Comparing the iteration numbers and calculation times of the Algorithms 4.4.1 and 4.4.2 with the numerical effort of the Algorithms 4.3.1 and 4.3.2 respectively we observe that the additional regularization term leads only to small extra costs. Taking into account that the regularized approaches promise higher stability they can be considered as valuable alternative to the accelerated Landsweber iteration.
- Again the further accelerated version leads to a considerable decrease of the necessary iteration numbers and the calculation time is still much lower despite the somewhat higher numerical effort in each iteration.

In a second experiment we try to verify the convergence rates results based on the stopping criterion (STOP2). Therefore we choose $\mathcal{X} = \mathcal{Y} = L^2(0,1)$. Moreover we consider the sample functions $x_1^\dagger$ and

$$x_3^\dagger(t) := -t^3 + \frac{3}{2}t^2 - \frac{19}{20}t + \frac{9}{20}, \qquad t \in [0,1].$$

Then $x_3^\dagger \in \mathcal{R}(A^*)$ and $x_1^\dagger \notin \mathcal{R}(A^*)$ holds. Moreover, from [44, Example 4] we have $x_1^\dagger \in \mathcal{R}\left((A^*A)^\mu\right)$ for all $0 < \mu < \frac{1}{4}$. Generalizing alternatively the calculation in Example 3.2.1 (which was done therein for $x^\dagger \equiv 1$) we can prove a distance function $d(R; x_1^\dagger) \leq C\, R^{-1}$,

	Algorithm 4.4.1			Algorithm 4.4.2		
δ_{rel}	$N(\delta, y^\delta)$	time (sec.)	$\frac{\Vert x^\delta_{N(\delta,y^\delta)} - x_1^\dagger\Vert}{\Vert x_1^\dagger\Vert}$	$N(\delta, y^\delta)$	time (sec.)	$\frac{\Vert x^\delta_{N(\delta,y^\delta)} - x_1^\dagger\Vert}{\Vert x_1^\dagger\Vert}$
0.05	256	0.36	0.3081	28	0.17	0.3109
0.01	603	0.65	0.1541	107	0.48	0.1630
10^{-3}	3882	3.94	0.0639	415	1.64	0.0646
10^{-4}	47934	48.59	0.0316	2378	8.39	0.0320
10^{-5}	192772	192.75	0.0063	12557	42.36	0.0061

Table 4.3: Calculation times and reconstruction error for sample function $x_1^\dagger$

	Algorithm 4.4.1			Algorithm 4.4.2		
δ_{rel}	$N(\delta, y^\delta)$	time (sec.)	$\frac{\Vert x^\delta_{N(\delta,y^\delta)} - x_2^\dagger\Vert}{\Vert x_2^\dagger\Vert}$	$N(\delta, y^\delta)$	time (sec.)	$\frac{\Vert x^\delta_{N(\delta,y^\delta)} - x_2^\dagger\Vert}{\Vert x_2^\dagger\Vert}$
0.05	482	0.59	0.3870	99	0.89	0.3839
0.01	2678	2.75	0.1663	248	1.16	0.1676
10^{-3}	24255	24.75	0.0464	1200	4.75	0.0483
10^{-4}	160417	161.47	0.0058	8268	29.12	0.0056
10^{-5}	896524	903.17	$5.819 \cdot 10^{-4}$	40812	141.73	$4.667 \cdot 10^{-4}$

Table 4.4: Calculation times and reconstruction error for sample function $x_2^\dagger$

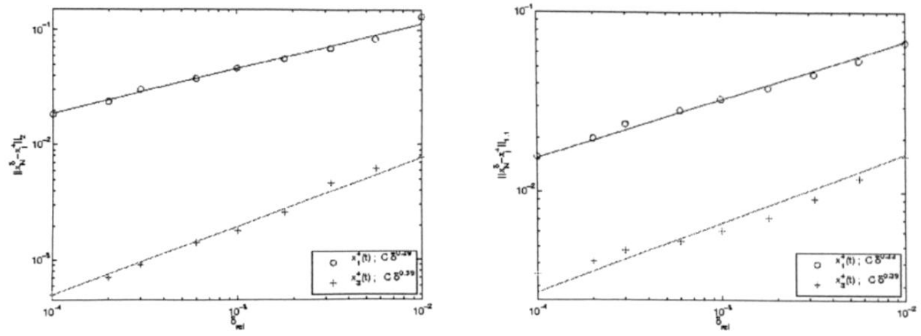

Figure 4.2: Convergence rates for $x_1^\dagger$ and $x_3^\dagger$ for $\mathcal{X} = L^2(0,1)$ (left plot) and $\mathcal{X} = L^{1.1}(0,1)$ (right plot)

$R \geq 1$, for some constant $C > 0$. This coincides with the corresponding source condition. Applying Theorem 4.4.3(ii) we choose $\bar{R} := C_{\xi^\dagger}\delta^{-\frac{1}{3}}$. For $x_3^\dagger$ we can set $\bar{R} \equiv const. = C_{\xi^\dagger}$ which is sufficiently large. The numerical results are presented in the left plot of Figure 4.2. Theoretically we expect the rates $\|x^\delta_{N(\delta,y^\delta)} - x_1^\dagger\| \leq C\,\delta^{\frac{1}{3}}$ for the first and $\|x^\delta_{N(\delta,y^\delta)} - x_3^\dagger\| \leq C\,\delta^{\frac{1}{2}}$ for the second sample function. As we can observe the numerical results are somewhat better in both cases.

In a second variant we choose again $\mathcal{X} = L^{1.1}(0,1)$. From Example 3.2.1 and the discussion in Section 3.6 we can expect a distance function $d(R;\xi^\dagger) \leq C\,R^{-\frac{2}{11}}$. This suggests the choice $\bar{R} := C_{\xi^\dagger}\delta^{-\frac{11}{15}}$ which gives an expected convergence rate $\|x^\delta_{N(\delta,y^\delta)} - x_1^\dagger\| \leq C\,\delta^{\frac{2}{15}}$. For $x_3^\dagger$ we can again choose $\bar{R} \equiv const. = C_{\xi^\dagger}$ which should lead also to the convergence rate $\|x^\delta_{N(\delta,y^\delta)} - x_3^\dagger\| \leq C\,\delta^{\frac{1}{2}}$. Considering the numerical results in the right plot of Figure 4.2 we see that the numerical convergence rate is better than predicted in the first but worse in the second example.

Appendix A

The Lepskij principle

We deal with the basic ideas and properties of the Lepskij- or balancing principle, see e.g. [64], [67] and [6]. This strategy has been well-established in the recent years since it is easy to implement and applicable under relatively weak technical assumptions. In particular, it can also be applied in Banach spaces.

We summarize the most important facts. The main idea of the balancing principle is based on the decomposition of the approximation error of regularized solutions into two parts which both depend on the regularization parameter α. We state the assumption in detail below.

Assumption A.0.1 *For each $0 < \alpha \le \alpha_{max}$ and given data y^δ let x_α^δ denotes any regularized solution of (3.2) satisfying*

$$\|x_\alpha^\delta - x^\dagger\| \le \psi(\alpha) + \phi(\alpha) \tag{A.1}$$

for a known non-increasing function $\psi(\alpha)$, which can depend on δ and an unknown non-decreasing (index) function $\phi(\alpha)$.

For given (sufficiently small) $\alpha_0 > 0$, a real number $q > 1$ and maximal index $j_{max} > 0$ we define the (finite) sequence

$$\{\alpha_j := q^j \alpha_0 \ : \ 0 \le j \le j_{max}\}. \tag{A.2}$$

The maximal index j_{max} is chosen such that $\alpha_{j_{max}} \le \alpha_{max}$. Then we can present the following a-posteriori choice of the regularization parameter α.

Definition A.0.1 (Lepskij-Principle) *Let the sequence $\{\alpha_j\}$ be defined by (A.2). We calculate solutions $\{x_{\alpha_j}^\delta\}$ of (3.4) and choose the regularization parameter $\alpha_L := \alpha_{j_L}$ such that*

$$j_L := \max\left\{j \le j_{max} \ : \ \|x_{\alpha_i}^\delta - x_{\alpha_j}^\delta\| \le 4\psi(\alpha_i) \ \forall i \le j\right\}. \tag{A.3}$$

Then $x_\alpha^\delta := x_{\alpha_L}^\delta$ is chosen as regularized solution of (3.2).

Now we can establish the theoretical main results of the balancing principle, see also [67, Proposition 2 and Corollary 1].

Proposition A.0.1 *Let $\alpha_0 > 0$ be chosen such that $\phi(\alpha_0) < \psi(\alpha_0)$, $\{\alpha_j\}$ is given by (A.2) and the index j_L satisfies (A.3). Moreover, define $\hat{j} := \max\{j : \phi(\alpha_j) \le \psi(\alpha_j)\}$ and $\tilde{j} := \max\{j : \|x_{\alpha_i}^\delta - x^\dagger\| \le \psi(\alpha_i),\ \forall i \le j\}$. Then, under Assumption A.0.1,*

$$j_L \ge \tilde{j} \ge \hat{j} \ge 0 \quad \text{and} \quad \|x_{\alpha_L}^\delta - x^\dagger\| \le 6\,\psi(\tilde{\alpha}) \le 6\,\psi(\hat{\alpha}).$$

If – in addition – there exists a constant $1 < D < \infty$ such that $\psi(\alpha_j) \le D\,\psi(\alpha_{j+1})$, $0 \le j < j_{max}$, then

$$\|x_{\alpha_L}^\delta - x^\dagger\| \le 6D \min\{\psi(\alpha_i) + \phi(\alpha_i),\ 0 \le j \le j_{max}\}.$$

We modify the estimate (A.1). For the derivation of the convergence rate result based on a-priori choice of the regularization parameter α we established estimates of the form

$$\|x_{\alpha_L}^\delta - x^\dagger\| \le C_1\tilde{\psi}(\alpha) + C_2\tilde{\phi}(\alpha) \tag{A.4}$$

with non-decreasing (known) function $\tilde{\psi}(\alpha)$ depending also on the noise level δ and non-decreasing function $\tilde{\phi}(\alpha)$ whcih depends on the unknown (approximate) source condition. In particular, Assumption A.0.1 is satisfied with $\psi(\alpha) = C_1\tilde{\psi}(\alpha)$ and $\phi(\alpha) = C_2\tilde{\phi}(\alpha)$. As a-priori parameter choice we have chosen $\alpha = f(\delta)$ such that $\tilde{\psi}(f(\delta)) = \tilde{\phi}(f(\delta))$ and obtained the estimate $\|x_\alpha^\delta - x^\dagger\| \le (C_1 + C_2)\tilde{\psi}(f(\delta))$ and the corresponding convergence rate

$$\|x_\alpha^\delta - x^\dagger\| = \mathcal{O}\left(\tilde{\psi}(f(\delta))\right) \quad \text{as} \quad \delta \to 0. \tag{A.5}$$

We will now show that we keep the convergence rate (A.5) when the regularization parameter is chosen by the balancing principle.

Corollary A.0.1 *Assume all conditions of Proposition A.0.1 to be satisfied and α_L is chosen by (A.3) and the estimate (A.4) is valid. Moreover, assume $\hat{j} < j_{max}$. Then the error bound*

$$\|x_{\alpha_L}^\delta - y^\delta\| \le 6D \max\{C_1, C_2\}\tilde{\psi}(f(\delta))$$

holds. In particular we obtain the convergence rate (A.5) with $x_\alpha^\delta = x_{\alpha_L}^\delta$.

Proof. We can apply Proposition A.0.1 with $\psi(\alpha) := C_1\tilde{\psi}(\alpha)$ and $\phi(\alpha) := C_1\tilde{\phi}(\alpha)$. We introduce the notation $\bar{\alpha} > 0$ which satisfies $C_1\tilde{\psi}(\bar{\alpha}) = C_2\tilde{\phi}(\bar{\alpha})$. Obviously $\bar{\alpha} \le \alpha_{\hat{j}} < \alpha_{\hat{j}+1} \le j_{max}$ holds. Hence, by monotonicity

$$\|x_{\alpha_L}^\delta - x^\dagger\| \le 6\,C_1\tilde{\psi}(\alpha_{\hat{j}}) \le 6D\,C_1\tilde{\psi}(\alpha_{\hat{j}+1}) \le 6D\,C_1\tilde{\psi}(\bar{\alpha}) = 6D\,C_2\tilde{\phi}(\bar{\alpha}).$$

Let $\alpha_* > 0$ satisfy $\tilde{\psi}(\alpha_*) = \tilde{\phi}(\alpha_*)$, i.e. $\alpha_* = f(\delta)$. Assume $C_1 \le C_2$. Then $C_1\tilde{\psi}(\alpha_*) \le C_2\tilde{\phi}(\alpha_*)$ which implies $\alpha_* \ge \bar{\alpha}$. Hence

$$\|x_{\alpha_L}^\delta - x^\dagger\| \le 6D\,C_2\tilde{\phi}(\bar{\alpha}) \le 6D\,C_2\tilde{\phi}(\alpha_*) = 6D\,C_2\tilde{\psi}(\alpha_*).$$

On the other hand, if $C_1 \ge C_2$ then $C_1\tilde{\psi}(\alpha_*) \ge C_2\tilde{\phi}(\alpha_*)$ and hence $\alpha_* \le \bar{\alpha}$ holds. This provides

$$\|x_{\alpha_L}^\delta - x^\dagger\| \le 6D\,C_1\tilde{\psi}(\bar{\alpha}) \le 6D\,C_1\tilde{\psi}(\alpha_*).$$

The proof is complete. ∎

Bibliography

[1] R. Acar and C.R. Vogel. Analysis of bounded variation penalty methods for ill-posed problems. *Inverse Problems*, 10:1217–1229, 1994.

[2] L. Ambrosio, N. Fusco, and D. Pallara. *Functions of Bounded Variation and Free Discontinuity Problems.* Oxford University press, New York, 2000.

[3] A.B. Bakushinskii and M.Y. Kokurin. *Iterative Methods for Approximate Solutions of Inverse Problems.* Springer, Dortrecht, 2004.

[4] A.B. Bakushinsky. Remarks on choosing a regularization parameter using the quasi-optimality and ratio criterion. *USSR Comp.Math.Math.Phys.*, 24(4):181–182, 1984.

[5] A.B. Bakushinsky. The problem of the convergence of iteratively regularized Gauss-Newton method. *Comput.Math.Math.Phys.*, 32:1353–1359, 1992.

[6] F. Bauer and T. Hohage. A Lepskij-type stopping rule for regularized Newton methods. *Inverse Problems*, 21:1979–1991, 2005.

[7] J. Baumeister. *Stable solution of Inverse Problems.* Vieweg Braunschweig, 1987.

[8] B. Blaschke, A. Neubauer, and O. Scherzer. On convergence rates for the iteratively regularized Gauss-Newton method. *IMA J.Numer.Anal.*, 17:421–436, 1997.

[9] T. Bonesky, K.S. Kazimierski, P. Maass, F. Schöpfer, and T. Schuster. Minimization of Tikhonov functionals in Banach spaces. *Abstract and Applied Analysis*, pages Article ID 192679, 19 pp. DOI:10.1155/2008/192679, 2008.

[10] K. Bredies and D.A. Lorenz. Iterated hard shrinkage for minimization problems with sparsity constraints. *SIAM J. Sci. Comput.*, 30:657–683, 2008.

[11] L.M. Bregman. The relaxation of finding the common points of convex sets and its application to the solution of problems in convex programming. *USSR Comput.Math.Math.Phys.*, 7:200–217, 1967.

[12] M. Burger and M. Osher. Convergence rates of convex variational regularization. *Inverse Problems*, 20:1411–1421, 2004.

[13] I. Cioranescu. *Geometry of Banach spaces, Duality Mappings and Nonlinear Problems.* Kluwer Dortrecht, 1990.

[14] I. Daubechies, M. Defrise, and C. De Mol. An iterative thresholding algorithm for linear inverse problems with a sparsity constraint. *Comm. Pure Appl. Math.*, 57:1413–1457, 2004.

[15] P. Deuflhard. *Newton Methods for Nonlinear Problems*, volume 35 of *Springer Series in Computational Mathematics*. Springer Berlin, 2004.

[16] J.C. Dunn. Global and asymptotic convergence rate estimates for a class of projected gradient processes. *SIAM Journal on Control and Optimization*, 19(3):368–400, 1981.

[17] D. Düvelmeyer, B. Hofmann, and M. Yamamoto. Range inclusions and approximative source conditions with general benchmark functions. *Num.Func.Anal.Optim.*, 28(11):1245–1261, 2007.

[18] A. Dvoretzky. Some results on convex bodies and Banach spaces. In *Proc.Int.Symp. on linear spaces, Jerusalem*, pages 123–160, 1961.

[19] P.P.B. Eggermont. Maximum entropy regularization for Fredholm integral equations of the first kind. *SIAM J.Math.Anal.*, 24:1557–1576, 1993.

[20] H.W. Engl. Discrepancy principles for Tikhonov regularization of ill-posed problems leading to optimal convergence rates. *Journal of Optimization Theory and Applications*, 52:209–215, 1987.

[21] H.W. Engl, M. Hanke, and A. Neubauer. *Regularization of Inverse Problems*. Kluwer Academic Publishers, Dortrecht, 1996.

[22] H.W. Engl, K. Kunisch, and A. Neubauer. Convergence rates for Tikhonov regularization of nonlinear ill-posed problems. *Inverse Problems*, 5:523–540, 1989.

[23] H.W. Engl and G. Landl. Maximum entropy regularization of nonlinear ill-posed problems. In V. Lakshmikhantam, editor, *Proceedings of the First World Congress of Nonlinear Analysts*, pages 513–525. Berlin: de Gruyter, 1996.

[24] H.W. Engl and O. Scherzer. Convergence rates results for iterative methods for solving nonlinear ill-posed problems. In D. Colton (ed.) et al, editor, *Surveys on solution methods for inverse problems*, pages 7–34. Springer, Wien, 2000.

[25] R. Griesse and D. Lorenz. A semismooth Newton method for Tikhonov functionals with sparsity constraints. *Inverse Problems*, 24:Article ID 035007, 19p., 2008.

[26] M. Hanke. Regularization with differential operators: an iterative approach. *Numer. Funct. Anal. Optim.*, 13:523–540, 1992.

[27] M. Hanke. A regularizing Levenberg-Marquardt scheme with application to groundwater filtration problems. *Inverse Problems*, 13:79–95, 1997.

[28] M. Hanke, A. Neubauer, and O. Scherzer. A convergence analysis of the Landweber iteration for nonlinear ill-posed problems. *Numer.Math.*, 72:21–27, 1995.

[29] M. Hanke and O. Scherzer. Inverse problems light: Numerical differentiation. *Am. Math. Mon.*, 108(6):512–521, 2001.

[30] O. Hanner. On the uniform convexity of L^p and l^p. *Ark.Mat*, 3:239–244, 1956.

[31] M. Hegland. Variable Hilbert scales and their interpolation inequalities with application to Tikhonov rgularization. *Appl. Anal.*, 59:207–223, 1995.

[32] T. Hein. Regularization of ill-posed problems in Banach spaces - approximative source conditions and convergence rates results. *Preprint TU Chemnitz, Faculty of Mathematics.*, 28, 2007.

[33] T. Hein. Convergence rates for regularization of ill-posed problems in Banach spaces by approximative source conditions. *Inverse Problems*, 24:Article ID 045007, 10p., 2008.

[34] T. Hein. Convergence rates results for recovering the volatility term structure including at-the-money options. *J.Inv.Ill-Posed Problems*, 17:350–374, 2009.

[35] T. Hein. On Tihkonov regularization in Banach spaces - optimal convergence rates. *Applicable Analysis*, 88:653–667, 2009.

[36] T. Hein. Regularization in Banach spaces - convergence rates by approximative source conditions. *J. Inv. Ill-Posed Problems*, 17:27–42, 2009.

[37] T. Hein. Tikhonov regularization in Banach spaces - improved convergence rates results. *Inverse Problems*, 25(3):Article ID 035002, 18 p., 2009.

[38] T. Hein. A unified approach for regularizing discretized linear ill-posed problems. *Mathematical Modelling and Analysis*, 14(4):449–464, 2009.

[39] T. Hein and B. Hofmann. On the nature of ill-posedness of an inverse problem arising in inverse option pricing. *Inverse Problems*, 19:1319–1338, 2003.

[40] T. Hein and B. Hofmann. Approximate source conditions for nonlinear ill-posed problems - chances and limitations. *Inverse Problems*, 25(3):Article ID 035003, 16 p., 2009.

[41] H. Heuser. *Funktionalanalysis.* B.G. Teubner, Stuttgart, 1992. (in German).

[42] B. Hofmann. Approximative source conditions in Tikhonov-Phillips regularization and consequences for inverse problems with multiplication operators. *Mathematical Methods in Applied Sciences*, 29:351–371, 2006.

[43] B. Hofmann. Interplay of source conditions and variational inequalities for nonlinear ill-posed problems. Preprint 2009-6, TU Chemnitz, 2009.

[44] B. Hofmann, D. Düvelmeyer, and K. Krumbiegel. Approximative source conditions in Tikhonov regularization - new analytical results and some numerical studies. *Mathematical Modelling and Analysis*, 10:41–56, 2006.

[45] B. Hofmann, B. Kaltenbacher, C. Pöschl, and O. Scherzer. A convergence rates result in Banach spaces with non-smooth operators. *Inverse Problems*, 23:987–1010, 2007.

[46] B. Hofmann and P. Mathé. Analysis of profile functions for general linear regularization methods. *SIAM J.Numer.Anal.*, 45(3):1122–1141, 2007.

[47] B. Hofmann and P. Mathé. How general are general source conditions? *Inverse Problems*, 24:Article ID 015009, 2008.

[48] B. Hofmann and O. Scherzer. Factors influencing the ill-posedness of nonlinear problems. *Inverse Problems*, 10:1277–1297, 1994.

[49] T. Hohage. Logarithmic convergence rates of the iteratively regularized Gauss-Newton method for an inverse potential and an inverse scattering problem. *Inverse Problems*, 13(5):1279–1299, 1997.

[50] T. Hohage. *Iterative Methods in Inverse Obstacel Scattering: Regularization Theory of Linear and Nonlinear Exonentially Ill-Posed Problems.* Phd thesis, University of Linz / Austria, 1999.

[51] B. Kaltenbacher. Some Newton-type methods for the regularization of nonlinear ill-posed problems. *Inverse Problems*, 13:729–753, 1997.

[52] B. Kaltenbacher. Regularization by projection with a posteriori discretization level choice for linear and nonlinear ill-posed problems. *Inverse Problems*, 16:1523–1539, 2000.

[53] B. Kaltenbacher, A. Neubauer, and O. Scherzer. *terative Regularization Methods for Nonlinear Ill-Posed Problems.* Walter de Gruyter, Berlin, 2008.

[54] B. Kaltenbacher, Frank Schöpfer, and Thomas Schuster. Iterative methods for nonlinear ill-posed problems in Banach spaces: convergence and application to paramter identification problems. *Inverse Problems*, 25:Article ID 065003 (19pp), 2009.

[55] L.W. Kantorowitsch and G.P. Akilow. *Funktionalanalysis in normierten Räumen.* Akademier Verlag Berlin, 1978. (in German).

[56] K.S. Kazimierski. Minimization of the Tihkonov functional in Banach spaces smooth and convex of power type by steepest descent in the dual. *Comput. Optim. Appl.*, 2008. accepted for publication.

[57] K.C. Kiwiel. Proximal minimization methods with generalized Bregman functions. *SIAM J.Control Optim.*, 35:1142–1168, 1997.

[58] J. Köhler and U. Tautenhahn. Error bounds for regularized solutions of nonlinear ill-posed problems. *J. Inv. Ill-Posed Problems*, 3:47–74, 1995.

[59] C. Kravaris and J.H. Seinfeld. Identification of parameters in distributed parameter systems by regularization. *SIAM J. Control Optim.*, 23:217–241, 1985.

[60] K. Kunisch and W. Ring. Regularization of nonlinear illposed problems with closed operators. *Numer. Funct. Anal. Optimization*, 14(3-4):389–404, 1993.

[61] Y.K. Kwok. *Mathematical Models of Financial Derivatives.* Springer, Singarpore, 1998.

[62] L. Landweber. An iteration formula for Fredholm integral equations of the first kind. *Amer.J.Math*, 73:615–624, 1951.

[63] P.D. Lax and A.N. Milgram. Parabolic equations. *Ann. Math. Stud.*, 33:167–190, 1954.

[64] O.V. Lepskij. A problem of adaptive estimation in Gaussian withe noise. *Teor.Veroyatn.Primen.*, 35:459–470, 1990.

[65] J. Lindenstrauss and L. Tzafriri. *Classical Banach spaces II.* Springer-Verlag New York/Berlin, 1979.

[66] J. Locker and P. Prenter. Regularization with differential operators I: General theory. *J. Math. Anal. Appl.*, 74:504–529, 1980.

[67] P. Mathé. The Lepskij principle revisited. *Inverse Problems*, 22:L11–L15, 2006.

[68] P. Mathé and S.V. Pereverzev. Discretization strategies for linear ill-posed problems in variable Hilbert scales. *Inverse Problems*, 19:1263–1277, 2003.

[69] P. Mathé and S.V. Pereverzev. Geometry of linear ill-posed problems in variable Hilbert scales. *Inverse Problems*, 19:789–803, 2003.

[70] D. Milman. On some criteria for the regularity of spaces of type (B). *C.R.(Doklady) Acad.Sci.U.R.S.S.*, 20:243–246, 1938.

[71] V.A. Morozov. On the solution of functional equations by the method of regularization. *Soviet. Math. Dokl.*, 7:414–417, 1966.

[72] V.A. Morozov. *Regularization methods for ill-posed problems.* CRC Press, 1993.

[73] M.Z. Nashed. *Generalized Inverses and Applications.* Academic Press, New York, 1976.

[74] F. Natterer. Regularisierung schlecht gestellter Probleme durch Projektionsverfahren. *Numer.Math.*, 28:329–341, 1977.

[75] F. Natterer. Error bounds for Tikhonov regularization in Hilbert scales. *Appla.Anal.*, 18:29–37, 1984.

[76] A. Neubauer. On converse and saturation results for regularization methods. In E. Schock, editor, *Beiträge zur Angewandten Mathematik und Informatik, Helmut Brakhage zu Ehren*, pages 262–270. Shaker, Aachen, 1994.

[77] A. Neubauer and O. Scherzer. A convergent rate result for steepest descent method and a minimal error method for the solution of nonlinear ill-posed problems. *ZAA*, 14:369–377, 1995.

[78] B.J. Pettis. A proof that every uniformly convex space is reflexive. *Duke Math. J.*, 5:249–253, 1939.

[79] D.L. Phillips. A technique for the numerical solution of certain integral equations of the first kind. *J. ACM*, 9:84–97, 19962.

[80] R. Plato and G. Vainniko. On the regularization of projection methods for solving ill-posed problems. *Numer.Math.*, 57:63–79, 1990.

[81] E. Resmerita. Regularization of ill-posed problems in Banach spaces: convergence rates. *Inverse Problems*, 21:1303–1314, 2005.

[82] E. Resmerita and O. Scherzer. Error estimates for non-quadratic regularization and the relation to enhancement. *Inverse Problems*, 22:801–814, 2006.

[83] A. Rieder. On convergence rates of inexact Newton regularizations. *Numer.Math.*, 88:347–365, 2001.

[84] A. Rieder. *Kein Probleme mit Inversen Problemen.* Vieweg, Wiesbaden, 2003. (in German).

[85] O. Scherzer. A convergence analysis of a method of steepes descent and a two-step algorithm for nonlinear ill-posed problems. *Numer.Func.Anal.Optim.*, 17:197–214, 1996.

[86] O. Scherzer. A modified Landweber iteration for solving parameter estimation problems. *Appl.Math.Optim.*, 38:45–68, 1998.

[87] O. Scherzer, M. Grasmair, H. Grossauer, M. Haltmeier, and F. Lenzen. *Variational methods in Imaging.* Springer Science+Business Media, LLC, 2009.

[88] F. Schöpfer, A.K. Louis, and T. Schuster. Nonlinear iterative methods for linear ill-posed problems in Banach spaces. *Inverse Problems*, 22:311–329, 2006.

[89] T.I. Seidman and C.R. Vogel. Well posedness and convergence of some regularisation methods for non-linear ill-posed problems. *Inverse Problems*, 5:227–238, 1989.

[90] U. Tautenhahn. *Regularisierungsverfahren für lineare und nichtlineare nichtkorrekte inverse Aufgaben.* postdoctoral lecture qualification, TU Chemnitz, 1993. (in German).

[91] U. Tautenhahn. Optimality for ill-posed problems under general source conditions. *Numer.Funct.Anal.Optim.*, 19(3):377–398, 1998.

[92] A.N. Tikhonov. On solving ill-posed problems and method of regularization. *Dokl.Acad.Nauk USSR*, 151:501–504, 1963.

[93] A.N. Tikhonov and V.Y. Arsenin. *Solutions of ill-posed problems.* Wiley, New York, 1977.

[94] D. Werner. *Funktionalanalysis.* Springer Berlin, 1997. (in German).

[95] Z.-B. Xu. Characteristic inequalities of uniformly L^p spaces and their applications. *Acta Math.Sinica*, 32:209–218, 1989. (in Chinese).

[96] Z.-B. Xu and G.F. Roach. Characteristic inequalities of uniformly convex and uniformly smooth Banach spaces. *J.Math.Anal.Appl.*, 157:189–210, 1991.

[97] E. Zeidler. *Vorlesungen über nichtlineare Funktionalanalysis II.* B.G.Teubner Leipzig, 1977. (in German).

[98] E. Zeidler. *Nonlinear Functional Analysis and its Applications III.* Springer-Verlag New York, 1985.